Land, Trees and Tenure

Proceedings of an International Workshop on Tenure Issues in Agroforestry

Nairobi, May 27-31, 1985

Sponsored by the Ford Foundation

edited by

John B. Raintree

ICRAF and the Land Tenure Center
Nairobi and Madison
1987

Cover photo by Gerardo Budowski

Published 1987

The Land Tenure Center
The University of Wisconsin
1300 University Avenue
Madison, Wisconsin, U.S.A.
and
The International Council
for Research in Agroforestry
P.O. Box 30677
Nairobi, Kenya

Printed in the United States of America

ISBN 0-934519-01-3

Dedicated to Jack C. Westoby, a forester who has always seen the people as well as the trees.

Contents

Asia

Latin America

Part 3. Working Group Reports and Plenary Discussions

Part 4. Postscript

Preface

This volume is the result of an international consultative workshop held in Nairobi in May 1985, organized by the International Council for Research in Agroforestry in collaboration with the Land Tenure Center of the University of Wisconsin and made possible by a grant from the Ford Foundation. The workshop was the second in a series of collaborative activities undertaken by ICRAF and LTC. The first resulted in *Trees and Tenure: An Annotated Bibliography for Agroforesters and Others*, compiled by Louise Fortmann, then consultant to ICRAF, and James Riddell of LTC. The third phase of collaboration, which is just beginning as these proceedings go to press, will involve applied research on tenure issues in technology-generating research projects with countries in ICRAF's Agroforestry Research Network for Africa (AFRENA).

The purpose of the workshop was to bring together an internationally recognized group of experts on tenure and agroforestry in order to assess the current state of knowledge on tenure issues in agroforestry and to identify priorities for research. In order to sharpen the focus of the workshop and make the most efficient use of the time together in Nairobi, five general background papers (Part 1 of these proceedings) were commissioned and sent to participants prior to the workshop. Other participants were asked to prepare brief position papers (Part 2) reflecting their experience with tenure issues in agroforestry and to come to the workshop ready to focus quickly on the key issues of a regional and cross-cutting nature (Part 3). Subsequent to the workshop the authors of two of the background papers consented to prepare a summary postscript on the main research priorities identified by the workshop (Part 4).

In the planning and organization of the workshop every effort was made to achieve balanced regional, institutional and disciplinary representation. In the end, although the regional balance was not as even as originally hoped, all of the regions had a reasonably adequate representation in the working groups. Over 50 participants and observers attended the workshop, including authors of papers from Ghana, Kenya, Nigeria, Sudan, Zaire, India, Indonesia, Nepal, the Philippines, Thailand, Costa Rica, Dominican Republic, Mexico, U.K., and U.S.A.

As can be seen from the list of registered participants at the end of this volume the majority came from universities, research institutes and ministries in developing countries. Among the international organizations represented at the workshop, were FAO, the Ford Foundation, the International Center for Integrated Mountation Development, the International Livestock Centre for Africa, the Overseas Development Council and the World Bank. Nairobi based representatives of the Rockefeller Foundation, UNEP and HABITAT also attended. The institutional heterogeneity of the participants was rivaled only by their disciplinary diversity. Among the disciplines present were agriculture, forestry, anthropology, sociology, geography, ecology, economics, political science, and law--representing a broad cross-section of research and development workers concerned with tenure issues in agroforestry.

To enhance the comparability of the papers and provide a minimal focus for the workshop deliberations, participants were requested to give special attention, among the other concerns identified by them, to six questions of practical importance to agroforestry: 1) What are the main tenure problems and how are they related to the adoption of agroforestry innovations? 2) What agroforestry systems show potential for solving or mitigating existing tenure problems? 3) How do these considerations relate to the problems of disadvantaged groups (e.g. women, the landless, etc.)? 5) What research is needed to clarify particular tenure issues in agroforestry? 6) How do the above considerations relate to the planning and implementation of research and development projects (methodological and institutional issues)? As expected, none of the contributions followed these guidelines religiously, but they may have served some purpose in defining the scope and practical intent of the workshop.

The papers and working group reports speak for themselves. As a whole, they present a heterogeneous mixture of analyses, solutions and, indeed, symptoms of the problems brought to light by the workshop. In its own way, each of the papers contributes to a clearer definition of the issues. Aside from unavoidable attempts to rectify the language, as editor of these proceedings I did as little tampering as possible with the content and tone of the articles, in order to pass on a faithful record of the event.

I should like to express my appreciation to John Bruce, my partner in the planning of this workshop, and to Louise Fortmann and Jim Riddell who made important contributions throughout the planning period. Thanks are also due to Stella Muasya for the original typing and to Jane Dennis-Collins for the final formatting of the manuscript, to Linda Olemoyoi and Roy Parker for their diligent editorial assistance, to Kadi Warner for preparation of the index and other invaluable editorial assistance, to Carol Dickerman for seeing the manuscript through the publication process, and to Michael Pierick for composition of the final laser-printed volume. A major vote of thanks is due to the authors of the papers and other participants, whose contributions are responsible for the success of the workshop, as well as to the Ford Foundation for making it possible. Needless to say, all credit for the merits of these proceedings go to the workshop participants and all blame for any textural shortcomings to the editor.

John B. Raintree
Editor/Workshop Coordinator

Part 1

General Background Papers

Land Tenure and Agroforestry: A Regional Overview

James C. Riddell

Land Tenure Center, University of Wisconsin-Madison

Introduction[1]

Concern with the worldwide loss of woody plants has spread from the experts who first sounded the alarm to the public at large. Writers such as Eckholm, Brown and others have done much to enhance this awareness. Less well appreciated is that the vulnerability of the world's natural tree crop is tied directly to the vulnerability of much of the world's food production system. For example, the conversion of tropical forest into grassland for the production of beef, the so-called "hamburger connection," has been well publicized in the media, but the fact that this phenomenon is associated with cattle raisers often following landless slash-and-burn cultivators who followed logging operations is more often overlooked. The problem can too easily be seen as a policy failure signaling the need for more legislation, rather than a systemic relationship between low productivity on existing insufficient soil resources and a growing but powerless population in search of land for minimal sustenance.

The immigrant farmers must turn to marginal soil resources, since the better endowed lands are closed to them. If the loss of woody plants is to be reversed, trees and agriculture will have to be combined in a way that will result in the more intensive use of existing landed resources--that is, some form of agroforestry. However, the use of trees to stabilize steep hillsides, erosion control, the meeting of fuelwood needs and so on, means that very real land tenure decisions must be made on who gets access to the improved resources and for how long. It is an area where we have not been very successful to date. There is ample documentation of participants in taungya, afforestation and woodlot projects tearing out or burning the very trees they have planted in order to continue using the land. Short of placing a guard for each tree, new institutional arrangements will have to be created.

Since we have all grown up with trees and within a land tenure system, much of our non-technical appreciation of either is largely culturally determined. The agroforester knows that each ecological situation calls for modification in tree selection, planting, weeding, spacing and so forth, and that one cannot simply choose one best tree/crop mix that will work everywhere. It is just the same with social systems and their arrangements over who has access to which parcels of land, for how long and for which purposes. This collection of practices, rules and social and institutional relationships resembles trees in that it is also environmentally specific--in this case specific to each sociocultural environment. As experience has shown, an otherwise technically sound agroforestry or forestry project can fail because it is viewed as a threat to existing land tenure rights and responsibilities. We need to look beyond the *de jure* land laws to the *de facto* "law in action" of day-to-day land tenure.

Land tenure as a "bundle of rights"

Land tenure, derived from the Latin *tenere* (to hold), refers to the possession or holding of the many rights associated with each parcel of land. We commonly refer to these rights associated with a parcel of land as a bundle that can be broken up, redivided, passed on to others and so forth. Some will be held by individuals, some by groups and others by political entities. For any land tenure system, each of the rights in the bundle will have at least three dimensions: people, time and space. We will examine each of these in turn and their relationship to trees.

People

No one anywhere ever owns land in a totally exclusive way. The rights of others always impinge to some degree on land use. Even where individual rights are most strongly developed, the society normally reserves to itself the right to take land for roads, and one cannot use one's land in such a way that it unfairly decreases the utility or value of a neighbor's land. And so it is not surprising that in these parts of the world a person whose orchard, ornamental or timber trees are diseased can be made to destroy them if they pose a danger to the trees of others. Indeed, this right of intervention is often most developed where concepts of private property are the strongest. This is just one of the many ironies associated with land tenure.

What it boils down to is that all of the rights that make up the bundle are the result of human interaction. They are an expression of social relationships and the expected behavior of others in response to actions I take *vis-a-vis* a piece of ground. If a forestry or agroforestry plan unwittingly increases the ambiguity of my relationships with others over that parcel of earth I depend on, the chances of my cooperation decrease in direct proportion to the insecurity created.

Time

It is in the dimension of time that the agroforester often encounters the most intensive interaction with a land tenure system. Since land survives any holder of rights in land, all social systems have mechanisms for the transfer of rights and for determining how long these rights last. Trees by their relatively slow maturation in comparison to crops and animals imply, if they are purposefully planted, that a party has the intention of possessing the land under the trees for an extended period.

For many parts of Africa and Asia, the community member who plants cash crop trees, such as rubber, cocoa or coffee on communal land, has effectively excluded all the other right holders from using "their" land for most other uses normal to their society. If an outsider plants trees, the land effectively passes to his or her kinship group. How much more reassuring is the prospect when the Forestry Department borrows land from the local community to plant trees? People have learned that where trees are concerned, what started out as a temporary transfer can all too easily become a long-term one. Indeed, the early failure of a forestry or agroforestry project, may resolve for the recipients the question of just which rights have been lost for how long.

Space

Each right in the bundle also defines just what are the spatial dimensions of use. The existing rights to space, however, can be quite complicated. What we find reported in the literature is that the same space is used for different purposes by different people at different times. The gum arabic (*Acacia senegalensis*) grove is a good example. Herders bring their animals into the groves to feed on the ground cover, the new seedlings and lower branches. Local farmers come to collect deadfall for firewood. Merchants purchase collection rights to the gum from the tree owners. The land upon which the trees grow is most often part of a large social group's communal holdings. In all likelihood, each of these users will seek to maintain their rights. As a general rule, the greater the change in the way land is used, the more people will have to be involved in acceptance of the idea.

Transfer

Each of the rights in the bundle has an exchange value in the society. The nature of the transfer of rights between persons or groups--just what is transferred, and to which kinds of people--makes up the majority of land tenure litigation and discussion in any society.

Different societies see the legitimate purposes of transfers quite differently, and set different limits upon transfers. In a classic statement of an African perspective, an Oil Rivers chief explained that "land belongs to a vast family, the ancestors, the living and the yet to be born." What the chief was saying is that according to traditional rules an individual can only enter into an exchange relationship for a piece of land for relatively short periods of time. Hence, loaning, borrowing, pledging and similar arrangements are common under communal tenure systems. A person cannot, however, alienate the land to a non-kinsman because of the impossibility of getting all those who hold rights in it together for the sale.

While this is only one example out of many variations cited in the literature, it does raise a warning flag for agroforesters. We need to find out whose rights are added to or detracted from by planting trees. Since land planted with commercial trees is usually more valuable than it was before, it is not surprising to find many traditional formalities governing tree planting.

What, for example, does the person who previously had rights of trespass and wood gathering on a parcel get in exchange when good management practices require more restrictive access? Unless there is an acceptable exchange that extinguishes these rights, persons in this category may be less receptive to ideas that have long-term advantages. Projects must determine to whom those advantages will accrue and arrange careful tradeoffs when the distribution of benefits is altered by the introduction of tree crops.

Africa

Land tenure in its *de facto* manifestation of social relationships on the land is always dynamic. Nowhere is this more evident than in Africa, where the need for agroforestry is the most pressing. In Africa trees have played a very important role in defining long-term use of communal land by individuals. This explains the rich literature that has grown up on tree planting and land tenure (see Elias 1962 or Benneh 1984, for example). Africa raises, however, important enigmas for the agroforester. In spite of rapid deforestation, resulting land deterioration, soil erosion and exhaustion, Africa contains over half of the world's remaining unutilized land resources. The continent also has the most equitably distributed land resources in the third world, yet almost all development efforts threaten this equality.

In the growing awareness of the magnitude of the emerging food crisis in Africa, agroforestry provides one of the few bright spots in the search for compatible technologies that will make existing land resources more productive for longer periods of time. Yet at the same time, the literature documents that the production of individual farmers is falling. Much current research points to policies that hold down producer prices and threaten traditional guarantees to land as major factors in this decline. If this is indeed true, then forestry and agroforestry policy must be seen by rural producers as increasing their control over their traditional resources if trees are to be effectively integrated into the effort to increase food production.

In a recent eight country study commissioned by FAO on changes taking place in land tenure in selected villages, the most startling result was the increasing importance of indirect tenures such as sharecropping (see Riddell 1984 for a summary). This bodes poorly for agroforestry. Trees and land have usually manifested separate tenure histories in Africa (e.g. Abrahams 1967, Cory 1970, Fortmann 1985). That is, people have rights concerning the trees they have planted, inherited, purchased, etc., even if the trees in question are on the land of another. For this reason, all across the continent there is a great reluctance to allow sharecroppers, renters and other indirect tenure holders to plant trees.

Trees have traditionally been a major factor in African tenure. Cocoa, oil palm, etc., quickly became integrated into local economies in the moist forest belt, with corresponding changes in prevailing land tenure systems. In the arid regions, colonial and post-independence emphasis on peanut and other non-woody plant cash crops has undercut the traditional gum groves, village wood lots and sacred stands (Sene 1980). Today, fuelwood is only slowly becoming seen as a cash crop. With the loss of local control over wood resources through legislation, such as the French A.O.F. Administrative decree of 1934 which made all natural forests state land, people have regarded fuel resources as open to all. In many regions fuel wood is not owned until it is cut and bundled (Ay 1978, Christiansson et al. 1981, Fisher 1953).

Agroforestry has much to build upon in the traditional African economy. Not only have trees been a means of establishing tenure, they also have cultural significance all across the continent. It is common to find trees left standing in cleared fields, or planted by houses as a repository for spiritual forces, family well-being and so forth.

Latin America

I would argue that this contrasts sharply with the role of trees in the agricultural life of Latin America. The two dominant problems facing the forests of Latin America are colonization of new lands and cattle ranching. In contrast to Africa, where trees are often planted to establish tenure rights, in Latin America they are often *removed* to demonstrate claims to land. Further complicating the work of the agroforester is the fact that the two processes of colonization and cattle ranching are so often intertwined (Hecht 1981). Colonists who open new land, whether as part of a government scheme or as a spontaneous response to landlessness, must first clear it to gain title. Within a short time the soil is exhausted and the farmers sell out to the large ranching enterprises which are just as anxious as the settlers to remove trees from their holdings (Denevan 1981).

The literature on agroforestry in Latin America has yet to address the issue of land tenure in anything approaching the detail evinced in the African and Asian material. This is no doubt associated with the fact that any changes in the use of the land involve major political economy issues. Those who are trying to form policy determining who gets rights to how much land for which purposes, must deal with a situation where, in many parts of Latin America, very powerful individuals or corporations posses an inordinate share of the soil resources (Hecht 1982, Cornick 1983).

Land reform has apparently exacerbated the situation rather than facilitated the incorporation of agroforestry into farming practices. In spite of the best of putative intentions, larger economic units occupy the better flat lands and the smallholders have cleared marginal hillsides, to establish their claims (Ampuero 1979). Much work is needed to develop agroforestry systems that are income enhancing on such small plots. For the land wealthy, deforestation for beef production is a viable alternative, even if its per-hectare return is lower than that achievable under agroforestry (see, for instance, Denevan 1982).

Political issues also dominate tenure policy in the remaining forest zones of Latin America. The basic civil rights of forest dwelling Indians need to be secured before land tenure legislation will have any effect (see Grasmick 1979 for a good discussion). Unlike Africa and Oceania, where traditional tenure rules form the negotiating base from which any change will emerge, the civil law in Latin America, as it has evolved from the colonial period, has not demonstrated any greater capacity than U.S. common law to protect and work with land tenure systems that have a non-western derivation (Ruddle 1974).

In light of the above, the Latin American farmer will need the security of a clear title if the threat of expropriation of improved rural resources by others is to be removed (Sellers 1977). A critical research issue that is not adequately addressed in the literature is the relationship between tree planting as a means of establishing some sort of more permanent claim to land, and tenure insecurity as a disincentive to planting trees at all (Seligson 1980). Agroforestry will have a long uphill fight toward widespread acceptance as long as tree destruction is the principle means of establishing rights in land (Schmithusen 1981). This contrasts with a situation that is just as bad from a forester's perspective but for different reasons in the South Pacific.

Oceania

Land tenure and agroforestry intersect along two distinct lines in Oceania (here broadly defined as Micronesia, Melanesia, Polynesia, Australia, New Zealand and the Malay Archipelago)[2]. The first has been population growth that has placed extreme pressure on all land resources and has made tree planting a difficult option. The second has been the conflict between communal and traditional concepts of land tenure and ideas of exclusive rights that accompany new economic uses (Crocombe 1964). This latter conflict goes back to the days of plantations and continues into the present "logging decades."

The literature is divided into two camps, one of which argues for modern land tenure and the other for protecting traditional land-owning groups under more modern legal forms. On those occasions when western laws have been imposed, it has often led to confusion and unwillingness to plant long-lived crops such as trees. Using the concept of the corporation by viewing traditional landholding groups such as lineages or tribes as legally constituted property holders has proven useful in the promotion of large-scale forestry in New Zealand (Williams 1979). Such incorporation has been found to facilitate group decision-making and reimbursement for long-term investments such as tree crops.

From the point of view of the forester who seeks unambiguous rights to large tracts of land, or of a government agency organizing a reforestation project, the conflict between modern and traditional systems of land tenure is a major source of concern in Oceania. It is perhaps to Asia that we must turn for the clearest examples of new social entities accepting the responsibility of introducing woody plants into an agricultural system in an effort to reverse processes of environmental deterioration.

Asia

Although integrated agroforestry is probably nowhere better illustrated than in the traditional agricultural systems of China and Japan (see King 1910 for an interesting appraisal by one of the founders of soil science), the land tenure issue most often associated with forestry and reforestation in South Asia is the "tragedy of the commons." The problem is simply that common lands, including forests, are owned by all and protected and maintained by none.

In India, for instance, many "panchayat" or communal lands have been overused for fuel and fodder for tens if not hundreds of years (Pant 1983). The responsibility for replanting and control of use is unclear in a changing economy (Shiva et al. 1983). The resultant environmental damage in India appears in the form of silted rivers, soil erosion and flooding. Similarly, India's forest cover is decreasing, as the combined impact of government-sponsored forestry and continued use by forest dwellers overload the forests' natural capacity to regenerate (Fernandes 1983).

In Nepal, Thailand, Malaysia and Burma the increase in both human and animal population, the "logging decades" and resettlement schemes have led to a rapid destruction of the forests which in turn has led to environmental deterioration on an alarming scale. Even the traditional right of farmers to collect fuel, fodder and wood for construction is now taking its toll on forest covers, and reforestation efforts have come into conflict with established behavior patterns.

In a related development in Pakistan, "shamlat" or communal lands are under *de facto* control of wealthy landowners, though they don't own them (Cernea 1981). Here the tragedy of the commons is resolved by expropriation by a wealthy minority (see also Romm 1981). Private tenure does not automatically resolve this overuse of the commons by the powerful. As Briscoe (1979) found in Bangladesh, private tenure leads to landlessness and sharecropping. With insecure tenure sharecroppers do not plant long-term crops, echoing similar situations in Africa discussed above.

Another problem mentioned in an overwhelming proportion of the Asian literature is associated with squatters and swidden cultivators. There are two factors. First, there is a very poor understanding of the operating system of land tenure among shifting cultivators by foresters decrying the situation, and second is the fact that until the security of tenure issue is faced and resolved to the satisfaction of cultivators, there is little chance for success (Dove 1983, Generalao 1978). These issues also raise another very important land tenure problem for this part of the developing world. That is, these swidden systems have grown out of an accurate perception of the problems associated with sustained production on forest soils in the tropics (Conklin 1957, 1961). Without good agroforestry techniques working hand-in-hand with traditional farming practices, there is little hope of a land tenure system emerging that is compatible with good forest management in local communities. The data, sparse though they are, seem to indicate that the more secure the tenure and the more real the earning power of swidden farmers, the more likely they are to settle and turn to more intensive cultivation techniques (Spencer 1977, Benge 1977).

China, Korea and Nepal have all been experimenting successfully with the development of local-level institutions. In China, farmers who plant trees to stabilize their hillside farms are given a tenure that is three times longer than that of others. Here, agroforestry clarifies tenure (Ross 1983). In Korea, the traditional cohesion of the village has been strengthened. The village council now decides on local wood production and can force private land-owners to participate (Openshaw and Moris 1979, Gregerson 1982). Similarly, in Nepal, the efforts to reconstitute the Panchayat as the resource-controlling group has been rewarded by the reestablishment of village woodlots (Fleming 1983, Bote 1982).

This is in contrast to the record of failures where government tried to treat participants in taungya systems as laborers. Once the trees became established they were burned, uprooted or cut in order for families to continue using the land, or to provide land for their maturing children (Boonkird 1978, Soerianegara 1982). Successful taungya depends upon clear and secure contractual rights and the development of a sense of community (Boonkird 1978, Goswani 1982). The very fact that successful taungya makes the formerly marginal land valuable also means increased government control. Unless strong institutional frameworks are established the greatest danger associated with "success" is loss of land.

General trends and conclusions

Within the multiplicity of conflicting claims commonly attached to the land, it is imperative for those seeking change to find out at which levels the control of land allocation is taking place. A difficulty noted in the literature arises in much of the third world where the state has exercised its sovereign power and has laid claim to vast portions of the nation's land, but does not actually make any day-to-day determinations as to its use. Land that is allocated to projects is often actually already possessed by others, and has been since time immemorial. As so many taungya projects have found out, tenure change has to be renegotiated with the local community, the kinship groups and the individual farmers.

Because tenure rules are the result of existing social relations they are always in a state of dynamic change. As social realities change, so do the interpretations of existing tenure rules, and new rules are created.

Throughout the developing world two countervailing trends are evident. In terms of production decisions, tenure is becoming more individualized, even on communally held lands. On the other hand, new governmental institutions, both at local and national levels, are showing themselves increasingly willing to assume tenurial roles in order to stimulate resource development, to protect natural resources such as forests, or to stimulate regional development by changing the use of land. Pure or pristine tenure types, such as freehold private tenure at one end of a continuum or strictly communal tenure at the other end, no longer exist if they ever did. Rather, it is the case everywhere that tenure forms are mixed. In almost all communal tenure regions, one finds private land-ownership and land sales. Also, as the literature on Kenya underlines, even where private tenure has the full backing of administrative policy and legal machinery, traditional communal forms persist in actual use and transfer practices (Odingo 1985).

Our conceptualization and understanding of the role of trees and woody plants in an overall rural economy has changed faster than have the social realities of land use and control. Forests that were formerly viewed as part of a state of nature are now seen as social property to be husbanded as part of a national patrimony. It has proved to be more than a problem of proper legislation, however. Legislation is usually not lacking (Schmithusen 1981, Stokes 1979). What is needed is a clarity of policy in regard to the interplay between land use needs, on the one hand, and an evolving institutional framework, on the other. Within the institutional relationships that make up land tenure, agroforestry will have an uphill battle until the growing of woody plants assures security as well as productivity.

Notes

1. Several of the ideas in this paper have appeared in the introduction to *Trees and Tenure: An Annotated Bibliography for Agroforesters and Others* (Fortmann and Riddell 1985).

2. I should like to acknowledge the help of Alexa Fraser in pulling together these strands from a very diverse literature on Oceania and Asia.

References

Abrahams, R.G. 1967. *The Peoples of the Greater Unyamwezi, Tanzania.* International African Institute. London.

Ampuero, E.P. 1979. Ecological aspects of agro-forestry in mountain zones: the Andean Region. In Trevor Chandler and David Spurgeon (eds). *International Cooperation in Agroforestry: Proceedings of an International Conference.* ICRAF. Nairobi.

Ay, Peter. 1978. Fuelwood and charcoal in the West African forest: Field research in Western Nigeria. In W.G. Morgan, R.P. Moss and

G.J.A. Ojo (eds). *Rural Energy Systems in the Humid Tropics*. United Nations University. Tokyo.

Benge, M.D. 1977. The private sector's contribution to reforestation. *Canopy* 3:4-5.

Benneh, G. 1984. Dynamics of land tenure and agrarian systems in Africa: Ghana case study. Unpublished paper commissioned by FAO. Rome.

Boonkird, Sa-Ard. 1978. Taungya system: its applications, ways and means of improvements in Thailand. In *Proceedings of the 7th World Forestry Congress*. IUFRO. Jakarta.

Bote, P. 1982. A model for forestry-based rural development. *Canopy* 8:14-16.

Briscoe, J. 1979. The political economy of energy use in rural Bangladesh. Mimeo.

Cernea, M. 1981. Land tenure systems and social implications of forestry development programs. Staff Working Paper No. 452. World Bank. Washington.

Christiansson, C., U. Goranson, and L. Lundgren. 1981. *Endarasha and Othaya: Land Use, Socio-Economic and Environment in Two Areas in the Central Kenya Highlands*. Department of Geography, University of Stockholm. Stockholm.

Conklin, H.C. 1954. An ethnoecological approach to shifting agriculture. *Transactions of the New York Academy of Sciences* 17(2): 133-142.

Conklin, H.C. 1957. *Hanunoo Agriculture*. FAO Forestry Development Paper No. 12. FAO. Rome.

Conklin, H.C. 1961. The study of shifting cultivation. *Current Anthropology* 2:27-61.

Conklin, H.C. 1969. An ethnoecological approach to shifting agriculture. In Andrew P. Vayda (ed). *Environment and Cultural Behavior*. Natural History Press. New York.

Cornick, T.R.V. 1983. The social organization of production in Quimag, Ecuador: A case study of small-farmer production systems in the highland Andes. Ph.D. thesis. Cornell University. Ithaca, N.Y.

Cory, H. 1970. *Sukuma Law and Custom.* Negro Universities Press. Westport, Connecticut.

Crocombe, R.G. 1964. *Land Tenure in the Cook Islands.* Oxford University Press. Melbourne.

Denevan, W.M. 1981. Swiddens and cattle versus forest: The imminent demise of the Amazon rain forest reexamined. *Studies in Third World Societies* 13:25-44.

Denevan, W.M. 1982. *Causes of Deforestation and Forest and Woodland Degradation in Tropical Latin America.* Report to the Office of Technology Assessment, Congress of the United States. Assessment of Technologies to Sustain Tropical Forest and Woodland Resources. U.S. Office of Technology Assessment. Washington.

Dove, M.R. 1983. Theories of swidden agriculture and the political economy of ignorance. *Agroforestry Systems* 1:85-99.

Elias, T.O. 1962. *The Nature of African Customary Law.* Manchester University Press. Manchester.

Fernandes, W. 1983. Towards a new forest policy: An introduction. In Walter Fernandes and Kulkarni Sharad (eds). *Towards a New Forest Policy: Peoples' Rights and Environmental Needs.* Indian Social Institute. New Delhi.

Fisher, J. 1953. *The Anatomy of Kikuyu Domesticity and Husbandry.* Department of Technical Cooperation. Nairobi.

Fleming, W.A. 1983. Phewa tal catchment program: Benefits and costs of forestry and soil conservation in Nepal. In Lawrence S. Hamilton (ed). *Forest and Watershed Development and Conservation in Asia and the Pacific.* Westview Press. Boulder, Colorado.

Fortmann, L. 1985. The tree tenure factor in agroforestry with particular reference to Africa. Unpublished Ms. Department of Forestry and Resource Management, University of California. Berkeley.

Fortmann, L. and J.C. Riddell. 1985. *Trees and Tenure: An Annotated Bibliography for Agroforesters and Others.* Land Tenure Center and ICRAF. Madison and Nairobi.

Generalao, M.L. 1978. Land use pattern and its relation to agroforestry. *Canopy* 4:7.

Goswani, P.C. 1982. Agro-forestry practices and prospects as a combined land use system. *Indian Forester* 108: 385-396.

Grasmick, J. 1979. Land and the forest-dwelling South American Indian: The role of national law. *Buffalo Law Review* 27:759-800.

Gregerson, H.M. 1982. *Village Forestry Development in the Republic of Korea.* FAO Document GCP/INT/347/SWE. FAO. Rome.

Hecht, S.B. 1981. Cattle ranching in the Amazon: analysis of a development strategy. Unpublished Ph.D. dissertation. University of California. Berkeley.

Hecht, S.B. 1982. Agroforestry in the Amazon basin: practice, theory and limits of a promising land use. In S.B. Hecht (ed). *Amazonia: Agricultural and Land Use Research.* CIAT. Cali, Colombia.

King, F.H. 1910. *Farmers of Forty Centuries.* University of Wisconsin Press. Madison.

Odingo, R.S. 1985. Land tenure study in the Nakuru, Kericho and Machakos Areas of the Kenya Highlands. FAO Special Studies, *The Dynamics of Land Tenure and of Agrarian Systems in Africa.* FAO. Rome.

Openshaw, K. and J. Moris. 1979. The socio-economics of agroforestry. In Trevor Chandler and David Spurgeon (eds). *International Cooperation in Agro-Forestry: Proceedings of an International Conference.* DSE/ICRAF. Nairobi.

Pant, M.M. 1983. Harvesting water from ravaged sinaliks in India. In Lawrence S. Hamilton (ed). *Forest and Watershed Development and Conservation in Asia and the Pacific.* Westview Press. Boulder, Colorado.

Poulsen, G. 1981. *Malawi: The Function of Trees in Small Farmer Production Systems.* FAO. Rome.

Riddell, J.C. 1984. Dynamics of land tenure and agrarian systems in Africa: A synthesis of findings. Unpublished Ms. Land Tenure Center, University of Wisconsin. Madison.

Romm, J. 1981. The uncultivated half of India: Parts I and II. *Indian Forester* 107:1-23, 69-85.

Ross, L. 1983. Obligatory tree planting: How great an innovation in implementation in post-Mao China. In *Joint Committee on Chinese Studies of the American Council of Learned Societies and the Social Science Research Council Workshop on Policy Implementation in the Post-Mao Era*. Purdue University. Lafayette, Indiana.

Ruddle, K. 1974. *The Yukpa Cultivation System: A study of Shifting Cultivation in Colombia and Venezuela.* University of California Press. Berkeley.

Schmithusen, F. 1981. Recent trends in legislation in developing countries. In *Proceedings of the 17th International Union of Forest Research Organizations World Congress*. Japanese IUFRO Congress Committee. Kyoto.

Seligson, M.A. 1980. *Peasants of Costa Rica and the Development of Agrarian Capitalism*. University of Wisconsin Press. Madison.

Sellers, S. 1977. *The Relationship Between Land Tenure and Agricultural Production in Tucurrique, Costa Rica.* Centro Agronomico Tropical de Investigacion y Ensenanza. Turrialba, Costa Rica.

Sene, E.H. 1980. "L'organisation fonciere dans la zone aride du Senegal et la gestion des ressources forestieres. *TORC*, 43(1284):14-18.

Shiva, V., H.C. Sharatchandra and J. Bandyopadhyay. 1983. The challenge of social forestry. In Walter Fernandes and Kulkarni Sharad (eds). *Towards a New Forest Policy: Peoples' Rights and Environmental Needs*. Indian Social Institute. New Delhi.

Soerianegara, I. 1982. Socio-economic aspects of forest resource management in Indonesia. In E.G. Hallsworth (ed). *Socio-Economic Efforts and Constraints in Tropical Forest Management.* Wiley. Chichester.

Spencer, J.E. 1977. *Shifting Cultivation in Southeast Asia*. University of California Press. Berkeley.

Stokes, J.S. 1979. *Forest Legislation Swaziland Part II: Revised Parts, Comments and Explanations*. Ministry of Agriculture. Mbabane, Swaziland.

Williams, E.W. 1979. Afforestation on Maori land in New Zealand. In K.R. Shepherd and H.V. Richter (eds). *Forestry in National Development: Production Systems, Conservation, Foreign Trade and Aid*, Development Studies Centre, Australian National University. Canberra.

Tree Tenure: An Analytical Framework for Agroforestry Projects

Louise Fortmann

Department of Forestry and Resource Management, University of California-Berkeley

Introduction[1]

Trees and land are not inseparable companions. The principle that trees and land can be sold separately has been widely reported (Cory 1970, Obi 1963, Weinstock 1983). Under Ottoman land law, the trees on Cyprus were owned separately from the land and "the owner of one could not sell his interest without giving a prior right of purchase to the owner of the other" (Meek 1968). Elsewhere certain uses of trees and tree products are permitted to all regardless of who owns the land on which the tree is growing (Ahn 1978, Schebesta 1973).

Since agroforestry depends on people's rights to plant and use trees, which in turn depend on the prevailing systems of land tenure and tree tenure, understanding these relationships is crucial to the design of agroforestry projects. This paper examines the components of tree tenure, the inter-relationships between tree and land tenure and their importance to the field of agroforestry.

Components of tree tenure

Tree tenure consists of a bundle of rights which may be held by different people at different times. Four major categories of rights make up this bundle: the right to own or inherit, the right to plant, the right to use, and the right of disposal.

The right to own or inherit trees

Murray (1982) has emphasized the importance to agroforestry projects of the right of the planter to own trees and completely control their usufruct. But in many systems the right to own may be vested in the community or a kinship group larger than the household (Obi 1963, Ollennu 1962). In some cases, trees may be owned even when the land on which they are grown cannot be (Tozzer 1941). In other cases, as in Niger, forest regulations do not permit individual ownership of trees (Seif el Din 1981). The laws of succession involving trees can be as complex as those involving land. It has been reported that in Sudan palm trees could be subject to complex fractional ownerships arising from the laws of inheritance (Leach 1919).

The right to plant trees

The right to plant trees, an obvious prerequisite for agroforestry, may be restricted. Particularly in Africa, planting a tree may give the planter rights over the land on which it is planted. The result is that planting trees may be used as a means to acquire or maintain rights to land. In the 1940's Basotho chiefs discouraged the planting of trees on arable land for this very reason (Meek 1968) and the prohibition has carried over into more recent times (Duncan 1960). Planting restrictions have been reported elsewhere as well (Meek 1968). It also has been reported that trees may be planted as visible evidence of a claim to land (Brokensha and Glazier 1973, Swanson 1979, Weinstock 1983). In Tanzania, planting permanent trees without permission on the land of another could be construed as "misbehavior" and constitute grounds for eviction (James 1967). In Zambia, tree planting by the colonial government was regarded as an attempt to seize land (Ng'andwe 1976).

The principle that the planting of trees establishes a claim to the land was weakened in Tanzanian litigation. In 1965 in *Iddi Juda Omari v. Issa Abdalla*, the Court held the customary law--"when a person occupies a land and plants sisal or trees of a permanent nature on the shamba (field), then the shamba belongs to him, no matter how long he stays without cultivating it"--to be "unreasonable in modern Tanzania" (James 1971). But in many places this principle continues to hold currency and affects the ability of people to participate in agroforestry projects.

In other cases, the planter does not obtain land rights but does retain ownership of the tree and with it the right of access to the tree. Such rights of access may encourage the planting of trees but can be extremely disruptive to other uses of the land.

The right to use trees and tree products

This may include:

1) The right to *gather* may include the right to collect deadfall or lop off dead branches or strip off hanging bark as well as the right to gather things growing on a tree such as fungus, insects, or birds'

nests. For example, a Peruvian agrarian cooperative allowed free gleaning of eucalyptus bark (Skar, Samanez & Cotarma, 1982).

2) The *use of the standing tree* may be necessary for curing hides or for hanging honey barrels. Among the Kamba, honey barrels were hung in trees on one's own land, on the land of another with permission, or in unoccupied trees in the commonage. The owner of the honey barrel then had the right to ensure that his bees remained undisturbed (Penwill 1951).

3) The right to *cut* all or part of a living tree for timber or building poles can be an important one. The cutting of living trees for fuelwood is generally frowned upon unless fuelwood is scarce.

4) The right to *harvest produce* such as fruit, nuts, and pods does not necessarily accrue to either the owner of the tree or the owner of the land. In parts of Nepal it has been reported that fruit is considered a common good (Bennett 1981).

5) The right to *use produce under the tree* may be less restricted than other rights. Leaves and twigs and even fallen fruit often may be collected from the ground under trees which are owned by others. Among the Semarang of Malaysia in the 1920's, fallen fruit was a common good even *inside* a private compound (Schebesta 1973).

The right to dispose of trees

This may include:

1) The right to *destroy* the tree by uprooting or chopping down individual trees or by clearing a section of forest.

2) The right to *lend* the use of the tree to someone else.

3) The right to *lease, mortgage or pledge* the tree. Tree pledging or leasing is widely practiced (Ollennu 1962, Panday 1976). Pledging is distinguished from mortgaging by the fact that the pledgee assumes control of the tree immediately rather than after a process of foreclosure.

4) The right to *give away or sell* the tree either together with or separate from the land. In Tanzania, (Abrahams 1967) reported the right to sell fruit trees as a recent innovation and (Meek 1968) noted that rights in mango and other trees could be bought and sold.

Who has what rights?

Four classes of right holders must be considered separately: the State, groups, households, and individuals within households.

The State

The State exercises its rights in three basic ways:

1) It may *regulate* (or attempt to regulate) the use of trees owned by others. Certain species may be declared "protected" no matter where they are located, to be used only with official permission (which often involves paying a bribe). This sort of legislation has been reported as detrimental to agroforestry projects in Haiti (Murray 1982) and Niger (Thomson 1981).

2) The State may *prohibit* or restrict the use of trees in forest reserves. This may be a point of serious conflict between the State and people living near and within forest reserves. A forest reserve may contain land that is essential to the economy of the surrounding area (Gupta 1982, Knowlton 1972, Meek 1968). The State may allow outsiders to utilize forest resources while prohibiting local people from doing so (Mishra and Tripathi 1978).

3) The State may *allow limited use* of forest reserve land. In many countries people are allowed use of forest reserves to satisfy basic needs for fuelwood, fodder, poles and minor forest products. The taungya system takes this a step further and actually allows farming on forest land in return for planting and caring for tree seedlings. But where landlessness is acute, the taungya farmer may ensure that the seedlings do not survive in order to remain on the land. Farmers may also plant tree crops in order to establish a more permanent claim to the land.

Groups

Groups need to be considered in three categories.

1) Groups may be *geographically defined*. All the residents of a certain village may be granted use rights while the rights of "strangers" are denied or restricted. The principle that only "local people of the neighborhood" could harvest palm trees (Stopford 1901) was upheld in a 1926 Nigerian court case (Elias 1971). Similar rules restricting the rights of "strangers" are found across Africa (Lloyd 1962, Ollennu 1962, Tanner 1960).

2) *Kin groups*, which frequently may have a specific territory under their control, may also have tree rights vested in them (Dove 1976).

3) *Non-kin groups* such as corporations or cooperatives may (usually with the acquiescence of the State) exercise rights over a particular area. Such rights as those granted logging companies may in fact conflict with the ancient rights of local people (Kunstadter 1980). The establishment of cooperatives on the old haciendas of Peru led to a redefinition of use rights, causing much community conflict, as cooperatives could restrict the right to gather fuelwood to their members and their families, disenfranchising non-members and widows of members of long standing (Skar, Samanez and Cotarma 1982).

Households

Households may have different tree rights which reflect differences in religion, caste, class, ethnic group or geographical origin (Briscoe 1979).

Individuals

Individuals within a household may also have different rights depending on gender, birth order, or status within the family.

Factors affecting who has what rights

Three general sets of factors affect who may exercise what rights, when and over which trees: the nature of the tree, the nature of the use, and the nature of the land tenure system.

The nature of the tree

Different rules may attach to the tree depending on whether it has been planted deliberately or whether it is self-propagated. This distinction is based on the principle that labor creates rights. The general rule of thumb is that wild or self-propagated trees are community property (Brokensha and Glazier 1973, Cory 1970, Crocombe 1964, Duncan 1960, Swanson 1979). Yet in some places, wild trees growing on private land were reported to be the property of the landowner (Lloyd 1962). Planted trees generally belong either to the planter or to the owner of the land.

The doctrine of labor creating rights may be extended to wild trees if labor is invested in making the tree usable. Thus, the pre-European Ojibwa recognized private rights to wild sugar maples that had been tapped (Herskovitz 1952).

The nature of the use

The rules which apply to subsistence use often differ from those which apply to commercial use. Trees which are used for subsistence, particularly those which are growing on a commons, are often free for use by all. In contrast, the use of trees for commercial purposes may be restricted to trees growing on the seller's property or may be forbidden altogether depending on the particular use (Brokensha and Njeru 1977, Pogucki 1970).

Fuelwood has traditionally been a common good although the rules governing its access may tighten as it becomes increasingly scarce. The use of trees serving community purposes may be restricted. Particular trees or groves of trees may be considered sacred or the dwelling place of spirits and thus be protected from use or cutting (Brokensha and Riley 1977, Conklin 1975). This protection often also serves a soil and water conservation function.

Trees have long performed a service role in the form of shade. Cutting down shade trees may be prohibited (Wagner 1970) or considered to indicate a lack of intelligence (Schapera 1943). In Botswana it was forbidden to cut trees in the immediate vicinity of a village (Schapera 1943). On the other hand, shade trees are sometimes cut down because they are thought to harbor snakes (E. Colson personal communication).

The nature of the land tenure system

Although tree and land tenure are distinct, each affects the other. As has been shown, planting trees can be used to establish *de facto* private ownership of land. On the other hand, rights in land also affect rights to trees. Where land tenure is communal and tree rights are strong, it would appear that tree planters are advantaged in their rights to trees. This may be particularly true in the case of shifting cultivation.

Several categories of right holders may be distinguished:

1) *Tree planters*: In many societies the person who plants a tree is the owner of that tree (Abrahams 1967, Conklin 1975, Mi & Shaosong 1980, Obi 1963, Spencer 1977). In contrast, where private rights to land are strong the strength of one's rights to trees may depend on the strength of one's rights to land. In this case, landowners are relatively advantaged while those with temporary or weak claims to land (tenants, including both sharecroppers and lessees, squatters, pledgees, mortgagees and women) may be relatively disadvantaged because of the possibility of using trees to establish a permanent claim to land, as discussed above.

2) *Landowners*: In Lesotho the court held that trees belong to the owner of the land whether or not they had been planted with the chief's permission (Duncan 1960). Among the Fanti of Ghana, the landowner had full rights to trees on the land and could not, "be improved out of this land" with trees planted by someone else (Sarbah 1968). In Botswana, trees could not be cut on the field of another without permission (Schapera 1943).

3) *Tenants*: Among the Baganda, a tenant was not allowed to sell trees of any sort. He could plant economic trees but the landlord was entitled to get a share of the proceeds and to become the owner of the trees upon the departure of the tenant (Haydon 1960).

Under the system in Ghana the tenant is obliged to maintain economic trees that are growing on the land but may not plant any himself. The owner has the right to the fruits of these trees; the tenant may harvest for his personal use, but never for sale. Without the permission of the landlord, he may not fell certain trees (Ollennu 1962).

In Nigeria, among the Ibo the sharecropper may plant a few economic trees, the number being determined by the number that a typical farmer in the community would be growing. Use rights to any trees planted by the owner remain with the owner throughout the sharecropping tenancy. However, the rights to use self-propagated trees is vested in the sharecropper, who also retains the inheritable rights to any economic trees which he has planted even after the tenancy ends (Obi 1963). Also among the Ibo, the lessor retains the rights to economic trees but the lessee is permitted to climb and prune them to allow cultivation of the ground beneath them. In practice this means stripping off nearly all the leaves and branches. The lessee can also cut down non-economic trees if it is "safer and more reasonable in the circumstances" than pruning them. The lessee however, does have to protect economic trees from fire. If a specific tree is leased, the lessee can not damage the tree (Obi 1963). Annual tenants among the Yoruba have neither the right to plant trees nor to use existing ones (Lloyd 1962).

In Fiji a lessee could not cut down trees without the consent of the lessor nor could he sell "any forest produce growing on the land" (Meek 1968).

4) *Borrowers* are under similar restrictions (Sellers 1977, Swanson 1979).

5) *Squatters* have strong reasons to try to acquire title to the land. In Costa Rica, squatters who hold rights of access and usufruct try to make their tenure more permanent by planting trees, as the law requires compensation for crops and improvements. Productive coffee, and peach palm trees are generally worth more than the land (Sellers 1977).

On the coast of Tanzania, squatters traditionally could plant trees (except cloves) and have rights over the use of the trees (Bailey 1965).

A particularly troublesome case of squatters and trees was found on the coast of Kenya where squatters worked for a small wage and paid rent for land to cultivate for themselves. With the introduction of cashews in 1937, the value of the cashew crop was far higher than the value of the rent. Eventually, an agreement was reached under which squatters could plant cashews in which they had inheritable rights, but the interests of the landowners were also protected. In eleven of the 296 arbitration cases resulting from evictions beginning in 1957/58, two awarded the trees to the squatters, two split them half-half, and seven awarded them to the landowner. Unfortunately, the legal principles were not specified in this study (Shambi 1972).

6) *Pledgees or mortgagees* possess rights based on the principle that the usufruct of the trees serves as the interest on the loan. Among the Yoruba, it was felt that the pledgee should have use of the trees for a minimum of two or three years in order to be fairly compensated (Lloyd 1962).

Under Ibo law the pledgee may not plant economic trees without the consent of the owner. If he does, the owner may destroy them when the pledge is redeemed. Further, since the land and the trees may be pledged separately from each other, obtaining pledge to the land does not automatically imply the right to use the trees (Obi 1963).

In Ghana, the pledgee (depending on the rights of the pledgor) has use rights in any economic trees, including the rights to fell or tap palm trees. He may also grow economic crops. Only if the pledgor redeems the land prematurely, must he bear any of the cost of such improvements (Ollennu 1962).

7) *Women* as a category do not always have restricted rights. Among the Ibo, women have very detailed rights to trees (Obi 1963). Tanner found that nearly as many women owned palm trees as did men on the Tanzanian coast, reflecting the influence of Islamic law. Meek (1968) found that women could inherit rubber land in Malaysia.

However, in societies where women are not permitted to own land, their rights in trees are often restricted. Where planting trees establishes rights to land, women may be forbidden to plant trees in order to prevent them from using this to obtain land. Women's rights to both trees and land may be a function of residence and marriage. Where marriage is not a stable institution, women's tree rights may be equally precarious.

The implications of tree tenure for agroforestry

In any agroforestry system tree tenure issues must be carefully examined to avoid the following problems.

Loss of rights may result from an agroforestry project as a consequence of a number of factors:

1) The project may disturb or destroy rights to other uses of the land or the trees on it. Hoskins (1980) has described a fuelwood project which involved the clearing of "useless" bush that was in fact the source of numerous products. Agroforestry practices such as alley cropping may involve loss of the right to use the space where the trees are planted.

2) Certain practices for cultivating and protecting trees may result in the loss of gathering rights.

3) When the value of trees is increased there is a tendency for both land and tree tenure to shift from communal to private holdings. The effect of privatization of land in Kenya was to reduce access rights to trees, and hence certain tree-dependent activities such as the production of honey declined (Brokensha and Njeru 1977). Similar effects may be expected to accompany other shifts in tenure.

In order to identify who may lose what rights as the result of a project it may be useful to construct a table such as the following. The users category should include whatever groups are relevant in a given project area. For each category of use, each group's probable gain or loss should be recorded. The loss in rights might be a result of expropriation of land for the project, rising market values of trees and tree-related products, or privatization of land. For example, the commercialization of fuelwood through the establishment of woodlots might lead to the loss of fuel gathering rights by landless households. In the final column a judgement should be made as to which of the groups are net gainers or net losers. This table would allow a project design team to determine what categories of people might be overall gainers or losers as a result of the project and what kinds of rights are in greatest need of protection.

Protection of the trees can be a problem. The ability to exclude others from the use of trees and tree products is essential if tree planters are to reap the benefit of their investment. In the Nepalese case cited above, tree owners had no right to exclude others from the use of their fruit trees for religious reasons. While one may have a legal right to prevent others from using resources including trees, in communities based on a system of reciprocal rights and obligations this is often very difficult to do. The personal or institutional capacity to enforce exclusionary rights may be very small indeed.

Protection is a matter of particular concern when tree planting is undertaken on wastelands. If there are clear communal rights to use such trees, either because they are growing on community land or because of the use to which they are put, or if there are ambiguous or unenforceable rights to exclude others from their use, then the project may be in considerable difficulty.

Certain categories of users may be unable to participate in the project because they do not have the right to plant or own trees. This is likely to be true of the landless, those with temporary claims to land, and women. In many places, these three categories singly or in combination will comprise the majority of the population. Thus, a project which does not take this into account may end up serving a relatively advantaged minority of the population, or the project may be destroyed by those who are excluded from it.

Because trees can be used to establish rights to land, it is necessary to monitor who is planting project trees where. Agroforestry projects can be used by private individuals to establish private claims to communal land (see for example Cernea 1981). Similarly, it is necessary to ensure that the community accepts the planting of trees on community land by or for the use of otherwise disenfranchised people.

Conclusion

There has been an understandable tendency to think of agroforestry in terms of the biological meaning of trees and in terms of land tenure. Yet it is also necessary to consider the social meanings of trees, including the norms of tree use and the long standing body of customary law dealing with tree tenure. Such considerations are essential to an understanding of the management of resource systems that include trees, and in designing agroforestry projects. Identifying who holds what tree tenure rights will help the project designer avoid the unintended destruction of existing rights, the exclusion of certain groups from project benefits, or the capture of the project by an elite for its own purposes.

The identification of existing rights should not be taken as the definitive answer to the problem of tree tenure. Tree tenure rights have been shown to shift over time -- often in the direction of privatization, as tree resources have become scarce or commercially valuable- often affecting land tenure as well (Millon 1955). Systems of tree tenure are no more static than other forms of social organization. Indeed, the systematic study of agroforestry, which has resulted in such refinements as the alley cropping system, may bring about new rights and new pressures on existing tenure systems. It is with these new directions that future research must be concerned.

Table 1. Assessing the effect of an agroforestry project on use rights.

	Twig Leaves	Fuelwood	Fruits/ nuts	Fodder	Use of Standing Tree	Timber Cons- truction	Other Subsis- tence	Commer- cial	Gain or Loss
By Gender:									
Men									
Women									
By Landed Status:									
Landless households									
Households with temporary land rights									
By Farm Size:									
Small farmers									
Large Farmers									
By Special Users:									
Grazers									
Forest Department									

Note: This table has been adapted from Kunstadter (1980) at the suggestion of Robert Chambers.

Acknowledgements

The literature research on which this paper is based was accomplished while the author was a consultant for the International Council for Research in Agroforestry. The helpful comments of John Bruce, Robert Chambers, Elizabeth Colson, John Raintree, Diane Rocheleau, Emery Roe, Jeff Romm, and James Thomson are gratefully acknowledged.

Notes

1. A modified version of this paper was published in *Agroforestry Systems* 2(2):229-251

References

Abrahams, R.G. 1967. *The Peoples of the Great Unyamwezi, Tanzania.* International African Institute. London.

Ahn, B.W. 1978. Village forestry in Korea. Eighth World Forestry Congress. IUFRO. Jakarta.

Bailey, A.P. 1965. Land tenure: Its sociological implications with specific reference to the Swahili speaking peoples of the East African Coast. Unpublished Masters thesis. University of London. London.

Bennett, L. 1981. The Parbatiya women of Bakundala. In *The Status of Women in Nepal.* Volume II Part 7. Center for Economic Development and Administration. Kathmandu.

Briscoe, J. 1979. The political economy of energy use in rural Bangladesh. Mimeo.

Brokensha, D. and J. Glazier. 1973. Land reform among the Mbeere of Central Kenya. *Africa* 43(3): 182-206.

Brokensha, D. and E.H.M. Njeru. 1977. Some consequences of land adjudication in Mbeere Division. IDS Working Paper No. 320. Institute of Development Studies. Nairobi.

Brokensha, D. and D. Riley. 1977. Vegetation changes in Mbeere Division, Embu. IDS Working Paper No. 319. Institute of Development Studies. Nairobi.

Cernea, M. 1981. Land tenure systems and social implications of forestry development programs. Staff Working Paper No. 452. World Bank. Washington.

Conklin, H.C. 1975. *Hanunoo Agriculture.* FAO Forestry Development Paper No. 12. Elliots Books. Northford, Connecticut.

Cory, H. 1970. *Sukuma Law and Custom.* Negro Universities Press. Westport, Connecticut.

Crocombe, R.G. 1964. *Land Tenure in the Cook Islands.* Oxford University Press. Melbourne.

Dove, M. 1976. Tree rights, tree-holding units and tree-using units among the Melaban Kantuq: the factors of scarce land, scarce labor and scarce knowledge in their evolution. Paper prepared for the 1976 Meetings of the American Anthropological Association.

Duncan, P. 1960. *Sotho Laws and Customs.* Oxford University Press. Capetown.

Elias, T.O. 1971. *Nigerian Land Law.* 4th edition. Sweet & Maxwell. London.

Elliot, C. 1982. The political economy of sewage: a case study from the Himalayas. *Mazingira* 6(4): 44-56.

Gupta, P.N. 1982. The effects of government policy in forest management in the Himalayan and Siwalik region of Uttar Pradesh, India. In E.G. Hallsworth (ed). *Socio-economic Effects and Constraints in Tropical Forest Management.* Wiley. Chichester.

Haydon, E.S. 1960. *Law and Justice in Buganda.* Buttersworth. London.

Herskovitz, M.J. 1952. *Economic Anthropology.* Knopf. New York.

Hoskins, M. 1980. Community forestry depends on women. *Unasylva* 32(130): 27-32.

James, R.W. 1967. Judicial developments in customary law:some reflections. *Journal of the Denning Law Society* 2(1):41-58.

James, R.W. 1971. *Land Tenure and Policy in Tanzania.* East African Literature Bureau. Nairobi.

Knowlton, C.S. 1972. Culture conflict and natural resources. In W.R. Burch Jr., N.H. Cheek Jr., and L. Taylor (eds). *Social Behavior, Natural Resources and the Environment.* Harper & Row. New York.

Kunstadter, P. 1980. Implications of socio-economic, demographic, and cultural changes for regional development in northern Thailand. In J.D Ives, S. Sabhasri, and P. Vorauri (eds). *Conservation and Development in Northern Thailand.* United Nations University. Tokyo.

Leach, T.A. 1919. Date trees in Haifa Province. *Sudan Notes and Records* 2:98-104.

Lloyd, P.C. 1962. *Yoruba Land Law.* Oxford University Press. London.

Meek, C.K. 1968. *Land Law and Custom in the Colonies.* 2nd edition. Frank Cass. London.

Mi, Lin and Shao-song Cai. 1980. A brief account of forestry development in the people's communes in Shao-Qing prefecture Guadong province, China. In *Report of the FA/SIDA Seminar on Forestry in Rural Community Development.* FAO. Rome.

Millon, R.F. 1955. Trade, tree cultivation and the development of private property in land. *American Anthropologist* 57:698-712.

Mishra, A. and S. Tripathi, 1978. *Chipko Movement.* Ghandi Peace Foundation. New Delhi.

Murray, G.F. 1982. Cash-cropping agroforestry: an anthropological approach to agricultural development in rural Haiti. Presented at the Wingspread Conference on Haiti: Present State and Future Prospects. Racine, Wisconsin.

Ng'andwe, C.O.M. 1976. African traditional land tenure and agricultural development: case study of the Kundu people in Jumbe. *African Social Research* 21:51-67.

Obi, S.N. Chinwuba. 1963. *The Ibo Law of Property.* Buttersworth. London.

Ollennu, N.A. 1962. *Principles of Customary Land Law in Ghana.* Sweet & Maxwell. London.

Panday, K.K. 1976. Importance of fodder trees and tree fodders in Nepal. Unpublished thesis. Federal Technical University. Zurich.

Penwill, D.J. 1951. *Kamba Customary Law. Notes Taken in the Machakos District of Kenya Colony.* Macmillan & Company. London.

Pogucki, R.J.H. 1970. Gold Coast land tenure. In E. Cotran and N. Rubin (eds). *Readings in African Law.* Vol.I. Frank Cass. London.

Sarbah, J.M. 1968. *Fanti Customary Laws.* 3rd Edition. Frank Cass & Co. Ltd. London.

Schapera, I. 1943. *Native Land Tenure in the Bechuanaland Protectorate.* The Lovedale Press. Capetown.

Schapera, I. 1955. *A Handbook of Tswana Law and Custom.* International African Institute. Oxford.

Schebesta, P. 1973. *Among the Forest Dwarves of Malaya.* Oxford University Press. London.

Seif el Din, A.G. 1981. Agroforestry practices in the dry regions. In L. Buck (ed). *Proceedings of the Kenya National Seminar on Agroforestry.* ICRAF. Nairobi.

Sellers, S. 1977. The relationship between land tenure and agricultural production in Tucurrique, Costa Rica. CATIE. Turrialba.

Shambi, M.M. 1972. The problem of land ownership and cashew nut claims in Malindi coastal belt. Unpublished Paper. Kilifi.

Skar, S.L., N.A. Samanez and S.G. Cotarma. 1982. Fuel availability, nutrition & women's work in highland Peru: three case studies from contrasting Andean communities. World Employment Programme Research WEP10/WP23. ILO. Geneva.

Spencer, J.E. 1977. *Shifting Cultivation in Southeast Asia.* University of California Press. Berkeley.

Stopford, J.G.B. 1901. Glimpses of native law in West Africa. *Journal of the African Society* 1(1):80-97.

Swanson, R.A. 1979. *Gourmantche Agriculture Part I: Land Tenure and Field Cultivation.* Integrated Rural Development Project Eastern ORD, B.A.E.F. Upper Volta Contract USAID 686-049-78.

Tanner, R. 1960. Land rights on the Tanganyika coast. *African Studies* XIX: 14-25.

Thomson, J.T. 1981. Public choice analysis of institutional constraints on firewood production strategies in the West African Sahel. In C.S. Russell and N.K. Nicholson (eds). *Public Choice and Rural Development.* Resources for the Future. Washington.

Tozzer, A.M. 1941. Landa's relacion de las cosas de Yucatan. Papers of the Peabody Museum of American Archeology and Ethnology. Harvard University. Cambridge.

Wagner, G. 1970. *The Bantu of Western Kenya with Special Reference to the Vugusi and Logoli.* Vol. II. Oxford University Press. London.

Weinstock, J.A. 1983. Kaharingan and the Luangan Dyaks: religion and identity in Central East Borneo. Unpublished Ph.D. dissertation. Cornell University. Ithaca.

Agroforestry, Tropical Land Use and Tenure

John B. Raintree

International Council for Research in Agroforestry

Introduction

Many people today have great expectations for agroforestry. Some of this optimism would seem to be justified on technological grounds, but if the current effort to understand, develop and disseminate agroforestry technology is to have any hope of meeting even a fraction of current expectations, the deployment of this newly organized branch of applied science must take place with a clearer than usual view of the human context of the supposed land use improvements.

The discussion of agroforestry is preceeded by an attempt to sketch a perspective on two of the most salient aspects of this context: 1) the current unprecedented rate of human population growth, and 2) existing patterns and trends in the development of tropical land use systems.

The purpose of this paper is to provide some mental images of the scope and potential role of agroforestry to serve as a background to the discussion of tenure issues. The main assumption behind the paper is that the interactions between agroforestry and tenure issues are basically of two types: on the one hand, tenure factors may pose constraints to the realization of the ecological and socioeconomic benefits of agroforestry and, on the other hand, agroforestry may offer ways of resolving or mitigating some of the existing tenure problems.

Without doubt, the tenure issues are far more varied and complex than what is reflected here. What I have tried to do is to focus attention on some of the major tenure changes that seem to arise as a regular concommitant of the main developmental trends in tropical land use, and to view them in ecological and evolutionary perspective. It is hoped that the workshop will elaborate on these patterns, bring out important divergences and additional complexities, illuminate the conditions under which they arise, and contribute to a sense of the priorities in dealing with them.

Human population: the regional context

The prodigious bloom of human population is, in evolutionary time perspective, a recent phenomenon which constitutes an unprecedented challenge to human adaptability as well as to the resilience of the earth's ecosystems. As Ansley Coale (1974) has remarked: "The present rate of world population increase--20 per 1,000--is almost certainly without precedent, and is hundreds of times greater than the rate that has been the norm for most of man's history. Without doubt this period of growth will be a transitory episode in the history of the population."

Although the consequences of this bloom in human numbers are debatable it is now generally acknowledged that the human population will grow to about 6.2 billion by the year 2000 and may reach 10 billion by the middle of the next century before there is any hope of stabilization.

If present trends continue

Africa is the only continental region that has experienced a consistent decline in per capita food production over the past 20 years, but Asia and Latin America are by no means assured of a smooth transition to the steady state. The FAO (1981) study, *Agriculture: Toward 2000*, has estimated the population that will be affected by serious undernutrition if present trends continue.

The largest growth in the number of seriously undernourished people is expected to occur in Africa, but the greatest number of affected people will be in Asia. Latin America experiences only a modest increase in the number of affected people in this scenario.

In the forestry sector, the study noted that fuelwood consumption in developing countries in 1979 was already some 100 million m^3 short of the estimated requirements and that an estimated 250 million people live in areas of fuelwood shortage. By the end of the century, due to the shrinking of the fuelwood resource base, the shortfall may rise to 1,100 million m^3, with some 3,000 million people facing acute fuelwood shortages. The nutritional and health consequences of inadequately cooked food could be serious (FAO 1981). To this must be added the deterioration of the productive potential of tropical land that accompanies the reduction of woody vegetation in the agricultural landscape.

Table 1. Incidence of serious undernutrition 1974/76 and 2000, by region. A calorie intake of less than 1.2 of the basal metabolic rate is taken as the criterion of serious undernutrition. Source: FAO (1981).

	Millions of Seriously Undernourished People	
	1974/76	2000
Far East	304	392
Near East	19	23
Africa	72	127
Latin America	41	46

Some of the main tenurial dimensions of these trends are succinctly summarized in the FAO (1981) study, as follows:

> The total number of agricultural households will grow very much faster than the arable land exploited in every region except Latin America. As a result, the average *size of holdings will decline* substantially in Africa and the Near East, and also in Asia. Accordingly, the number of *landless* agricultural laborers will increase, partly because the area of land held by some individuals or families will become too small to be viable and partly because large landholders will buy up the land of economically weak farmers.
>
> Because ownership of or access to farmland is usually the main influence on incomes of agricultural people, an inevitable outcome of unmodified trends would be a worsening of the present highly *unequal distribution of production assets and incomes.* Such trends aggravate social tension in rural areas, and conflict with objectives of greater equity in social and economic relations which are coming increasingly to the fore.

All of this adds up to a rather bleak picture of the human condition at the turn of the century, if present trends continue.

Alternative futures

While there is a certain demographic inevitability in the continued rise of population before it eventually levels off, no one really expects that the projected trends will continue unmodified. Man is, after all, a highly adaptive creature. Possessed of what Malinowski called "the curse of foresight and imagination," the shape of mankind's future is, to a very large extent, a question of *effort*.

Agriculture: Toward 2000 (FAO 1981) advances two rather ambitious change scenerios which, if realized, would substantially mitigate the effects of population growth in the developing world. Both of these scenarios envisage increases in the level of inputs to agriculture which recent history gives us little encouragement to believe possible. Nevertheless, the direction of the needed changes in land use are reasonably clear.

Where will the production increases come from?

There are basically three possibilities: 1) bringing more land into cultivation, 2) increasing the frequency of cropping on existing agricultural lands (i.e. intensification of shifting cultivation), and 3) increasing the yields per unit of cultivated land (technological intensification). The projected regional contributions from each of these sources are shown in Table 2.

Table 2. Contributions to growth in agricultural output, by region (Scenario A). Source: FAO (1981).

	Expansion of Land in Cultivation	Increase in Cropping Intensity	Increase in Yields per Unit of Land
Asia	10%	14%	76%
Africa	27%	22%	51%
Latin America	55%	14%	31%

In Asia, where some 79 percent of the arable land is already in cultivation, most of the production increase will have to come from technological intensification of agriculture resulting in higher yields per unit of land, although there is modest scope for intensification of shifting cultivation and expansion into new agricultural land. Substantial proportions of the targeted increases in Africa may come from expansion of cultivated (in most cases marginal) land and intensification of shifting cultivation, but the major part of the projected growth must come from increases in the yields per unit of land. Latin America, with a mere 25 percent of its arable land in cultivation as of 1980, has the greatest hypothetical potential for expansion of the area under cultivation. However, much of the projected expansion would require conversion of the main global reserves of tropical rainforest to agricultural uses of doubtful sustainability. In any event, a hefty increase in the productivity of agricultural land will also be required in the Latin American region.

Tropical land use systems: The ecological context

How do the potential sources of production increase relate to the existing patterns of land use in the tropics? What tenure problems crop up within and between different land use systems? A brief overview of tropical land use systems and their main tenure problems in this section will set the stage for the consideration of agroforestry potentials and tenure problems in the final section of this paper. For the sake of brevity, the focus is restricted to the tropics, where most of the developing countries are situated, although the perspective could probably be generalized to the subtropics without many modifications.

Figure 1 gives a broad overview of the major systems of tropical land use. More complete classifications of tropical land use systems have been published (Allan 1965, Gourou 1966, Ruthenberg 1971, Benneh 1972, Watters 1960, Spencer 1966, Unesco/UNEP/FAO 1978, 1979) but most of the subclasses are implicit in Figure 1. The intention, is not to be exhaustive, but rather to highlight in broad outline the salient characteristics of the most pervasive systems.

Table 3 relates the development of the different sources of production increase to the kinds of tenure conflicts which are most likely to arise.

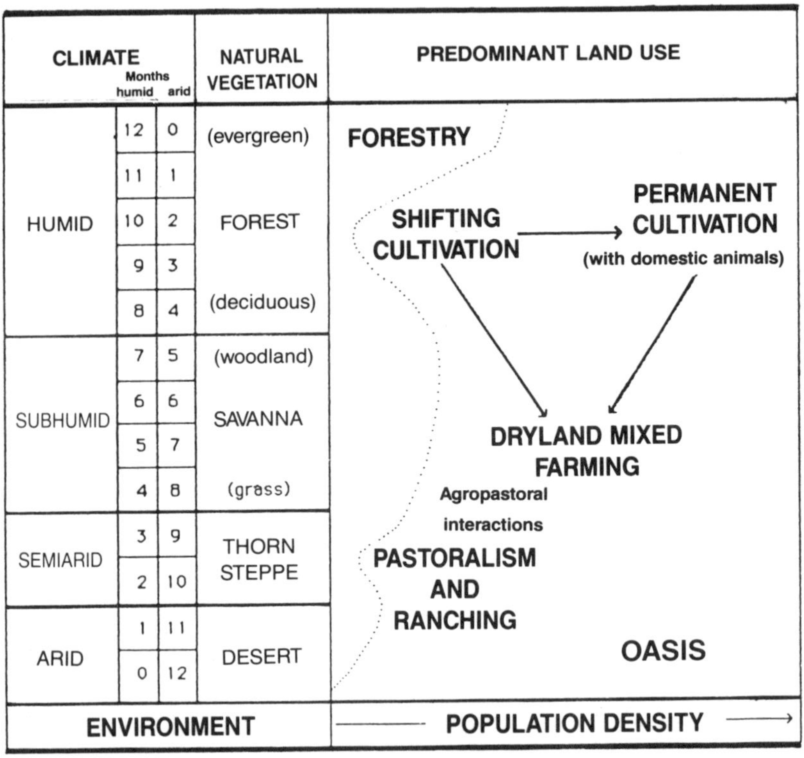

Figure 1. The human ecology of tropical land use, showing the predominant land use systems and development pathways as a function of environment and population density. The boundaries of the different systems are only approximate.

Table 3. Major tenure issues in relation to the potential sources of production increase.

Sources of Production Increase	Tenure Issues
1. Expansion of cultivation	
1.1 Expansion of shifting cultivation into reserved forests	shifting cultivators *vs.* foresters
1.2 Expansion of dryland farming into grazing lands	farmers *vs.* pastoralists, local grazers & gatherers
1.3 Expansion of farming onto sloping lands	farmers *vs.* foresters, local gatherers & grazers
2. Increase in the frequency of cropping	
2.1 Intensification of shifting cultivation (shortening of fallow period)	common *vs.* household rights to fallow land
3. Increase in yields per unit of land	
3.1 Technological intensification of shifting cultivation	common *vs.* household rights to improved land
3.2 Technological intensification of permanent farming	gender and inter-household conflicts, landlessness
3.3 Tree crop alternatives	long term *vs.* short term rights in land, tree tenure

Expansion of shifting cultivation into forest reserves

In the humid tropics the main source of tenure conflict between farmers and foresters is the expansion of shifting cultivation into reserved forests, either through spontaneous "enroachment" on forests by shifting cultivators in search of land to cultivate or through the officially sanctioned modification of shifting cultivation known as "taungya." In the former case, the forest land in question is subject to conflicting tenure claims: the longstanding traditional claim of the shifting cultivators to tribal land *vs.* the modern legal claim of the state to rights of forest exploitation. For traditional shifting cultivators this conflict may arise even at low levels of population density before land pressure has become serious. A second and more clearly defined case of "enroachment" arises in connection with land hunger among more settled agriculturalists, who deliberately enter what they acknowledge to be reserved forest in search of land.

One of the main forestry responses has been the "taungya" system of plantation establishment, whereby farmers are given temporary access to land for cultivation and sometimes a modest wage in return for their labor in the planting and aftercare of commercial forest seedlings. Kio (1972) has defined taungya as the "method of employing shifting cultivators to raise forest trees" (quoted in Enabor 1974). Enlightened variants of the taungya approach do exist and are discussed in the next section, but the tenure arrangements of classical taungya tend to bear a remarkable similarity to classical landlord-tenant relations. In this case the feudal landlord is replaced by the state forestry agency:

> The system is frankly exploitative in concept and operation and cashes in upon the needs of the landless and poor people to serve its own ends. The much vaunted incentives are only a cloak for uninhibited exploitation, as the savings effected by the Forest Department are many times more than the expenditure incurred on elementary conveniences provided to the working force (Seth 1981, quoted in Foley and Barnard 1984).

Foresters have enumerated the conditions necessary for the successful introduction of taungya as follows: 1) a low level of income or standard of living, 2) land hunger among the farming population, and 3) unemployment or underemployment. "These factors constitute important elements of the so-called cycle of poverty and may, to some degree, be attributed to the practice of shifting cultivation" (Enabor 1974). The same author sums up the classical forester's view of taungya as an "excellent opportunity for the maximum utilization of land resources, while at the same time affecting the desired control of traditional shifting cultivation."

Basically, then, classical taungya is a way of reducing the labor costs of plantation establishment or, in other words, of maintaining a supply of cheap, unskilled labor by perpetuating the "cycle of poverty" it pretends to address. It depends for its limited appeal to farmers on effective restriction of access to forest land. Boserup (1965) notes that in feudal Europe restriction of forest clearance, which forced the farmers to work in the landlord's fields, was an expedient in times of labor scarcity for the landlord. I have argued elsewhere (Raintree 1984) that conventional forestry is limited less by land scarcity than by a scarcity of skilled and committed labor. Farmers vastly outnumber foresters and, as long as farmers are viewed as at worst "encroachers" and at best "cheap labor," the traditional antagonism between the two will continue, to the ultimate detriment of sustainable forest production.

No doubt some accommodation between farmers and foresters is needed in land pressured areas where forestry agencies have the responsibility of producing timber and other forest products to meet national needs, but it is unlikely that unmodified forms of classical taungya will provide the answer. A more promising land use alternative would be one that allows farmers to become equal partners in the opportunities and responsibilities of a more integrated land use system. An example of such an approach is the Thai "Forest Village" scheme discussed in the final section of this paper.

Expansion of dryland farming into grazing lands

Traditional dryland farming in the developing countries offers at best an ecologically precarious existence. Plagued by crop failures due to drought or unpredicable rains, exhaustion of fragile soils, dry season feed shortages for draft animals, and difficulties in maintaining a sustained supply of fuelwood due to the slow regenerative capacity of natural woody vegetation, the system is nevertheless expanding rapidly as population expands into agriculturally marginal lands.

This expansion is associated with at least two distinct types of tenure conflict: 1) Conflict between farmers and non-farming pastoralists over the loss of grazing land and, from the other side, crop damage by herd animals. The former is particularly significant to pastoralists, since at the frontier of agricultural expansion it is usually the best grazing lands (often the critical dry season grazing reserves) that are the first choice of the farmers, while the latter is an important constraint on tree planting. 2) Conflict within the mixed farming community itself over the loss of grazing land for domestic animals and, as population density increases, the restriction of traditional gathering rights to fuelwood and other natural products.

Under pressure from the advancing agricultural frontier, pastoralists have been the focus of programs aimed at increasing the security of their tenure over pastoral lands. But the magnitude of the social learning process necessary to institutionalize the land use changes implicit in these schemes is easily underestimated. The Maasai Group Ranch Scheme in Kenya is an interesting case in point. Although the establishment of legal boundaries around the group ranches has succeeded in giving the Maasai a measure of freedom from land grabbing farmers and speculators, after more than a decade it is now fairly clear that the group ranches have not been able to successfully replace the traditional pastoral strategy with the social institutions necessary for common property management within the boundaries of the ranches. There is now a vigorous movement in the group ranches toward subdivision and individualization, which may have the ultimate effect of exposing individual ranchers, once again, to the risk of losing their land.

Table 4 provides an overview of agropastoral interactions in semi-arid areas of Africa.

Expansion of farming onto sloping lands

Hill farming is nothing new in the tropics, but steeper and steeper slopes are now coming into cultivation under pressure of population growth. The main farming hazard, of course, is that the topsoil under cultivation will be washed away. The problem is most severe in systems of continuous cropping, but is also quite significant with shifting cultivation under conditions of shortening fallow.

In hill areas of high population density the establishment of more ecologically appropriate and sustainable perennial crop systems is often hampered by small farm sizes. With barely sufficient land for subsistence food production, most smallholders cannot afford to forego immediate production in order to make the transition to the higher yielding and more sustainable tree crop systems. In many areas the severity of land scarcity is artificial, being associated with highly unequal landholdings.

Table 4. Agropastoral interactions in semi-arid regions of Africa. After McGown et al. (1983).

Types of Linkage	Comments
Between Different Management Units	
1. No linkage	Pastoralists and farmers operate as independent units; e.g. traditional relations between Maasai and Kikuyu in Kenya (now changed to exchange and competition linkages).
2. Ecological linkage	Crop residues grazed, manure deposited (may become exchange linkage).
3. Exchange linkage	Exchange of agricultural and livestock food products; crop residue-manure exchange with payment (direction of payment depends on relative scarcity).
4. Competition linkage	Competition for land between farmers and pastoralists (greatest during the dry season); land taken for cultivation usually the best grazing land; crop damage by livestock incursions; beneficial crop residue-manure linkage may mitigate the conflict somewhat.
Within the Same Management Unit	
1. Investment linkage	Livestock investments generally give higher returns than farming investments (most typical form of integration of livestock with farming in Africa); for capital development and emergency savings.
2. Food linkage	Cattle or more commonly smallstock kept for daily milk and occasional meat consumption.
3. Manure linkage	Critical to permanent cultivation systems; where field types of different intensity exist, manure usually concentrated on intensive near fields.
4. Draft linkage	Critical component of dryland mixed farming systems (constrained by low nutritional status of draft animals at beginning of rains).
5. Fodder linkage	Crop residues and fodder crops as supplementary dry season feed; in high density areas households with small number of domestic animals may practice cut-and-carry pen feeding (constrained by low quality of herbaceous feeds in dry season).

The "minifundia-latifundia" disparity in Latin America is perhaps the classic case in point, but similar problems are found wherever stratified communities exist, often in association with commercialized plantation agriculture. As demonstrated by a recent case study in the Philippines (Torres and Raintree 1983), the situation may present difficult technological challenges. In one part of the study area under traditional tenure, a highly productive and sustainable coconut-based multistorey tree crop system was expanding by means of "hacienda taungya", using tenants as the labor force for plantation establishment. In another part of the area coming under a government land reform program, tenant farmers were being given the opportunity to obtain title to the fields they had cultivated as tenants. The small size (usually under 2 ha) of these holdings, however, prevented these farmers from participating in the tree crop system for the previously mentioned reasons.

The only alternative known to the local smallholders was to continue the destructive plow cultivation technology which had been diagnosed as the main cause of erosion and gradually declining yields. Confronted with the paradox of "sustainability without equity" *vs.* "equity without sustainability," the agroforestry design team that conducted the study proposed an "alley cropping" system, combining aspects of the tree crop system with the traditional field practice. This case illustrates the role of technological innovations in support of social reforms.

Increase in the frequency of land use

The history of farming in the humid, and to a lesser extent the subhumid, tropics is by-and-large the history of agricultural intensification from shifting cultivation to permanent arable farming.[1] The basic outline of this evolutionary process is summarized in Table 5.

In stages 1 and 2, as population density increases, the intensification of shifting cultivation (swidden) takes the form, initally, of an increase in the frequency with which a given plot is cultivated, with an accompanying decrease in the fallow period. According to Boserup's (1965) analysis, this process is generally accompanied by a shift in traditional land tenure from common "tribal rights" to the gradual recognition of increasingly exclusive "family rights" to reuse of fallow land. (Note: "family" in the sense used here is differently defined in different societies.) At the forest fallow stage, every family enjoys common rights of access to swidden land by virtue of its membership in the tribe or other territorial unit. In some systems kin group heads or local chiefs may play a redistributive role, allocating land to household production units from the common pool. Since good land is not scarce, fallow land reverts to the common pool and the right to recultivate a given plot is generally only a matter of "courtesy" in egalitarian societies or a discretionary privilege in redistributive systems. As land use intensifies, however, family rights to the plot tend to become more fixed and exclusive, by dint of previous use and labor invested in clearing, although these rights may lapse through nonreuse.

Table 5. The main stages in intensification in the evolution of tropical agriculture from shifting to continuous cultivation. Source: Raintree and Warner (1985), after Greenland (1974) and Boserup (1981).

Stage 1	Stage 2	Stage 3	Stage 4
Simple shifting cultivation	Recurrent cultivation long term medium short	Recurrent cultivation with continuously cultivated plots shifting cultivation with permanent tree or other cash crops	Continuous cultivation
(Greenland 1974)			
Forest fallow	Bush fallow	Short fallow	Annual cropping Multicropping
Population density			
Very sparse 0-4/km^2	sparse 4-16/km^2	medium 16-64/km^2	dense 64-256/km^2 to very dense 256+/km^2
(Boserup 1981)			

At the bush fallow stage, when the better cultivation sites are becoming scarce, there is greater interest in recultivating a given plot and family tenure tends to become more exclusive. When the attachment to a plot becomes permanent, the custom of "pledging" may emerge, which involves the transfer of cultivation rights in the plot to another family, subject to being returned on request or after a contractually specified period.

> Thus, the attachment of individual families to particular plots becomes more and more important with the gradual shorting of the period of fallow and the reduction of the part of the territory which is not used in the rotation. By the same token, the general right of the members of the tribe to clear a new plot becomes less valuable, because the land which is not reserved fallow land for some family is now likely to be of inferior quality or very difficult to clear (Boserup 1965).

It is quite probable that in the history of any given area, it is precisely this "inferior" land which becomes the object of common grazing and gathering rights which become so important in the later stages of intensification. In this connection it may be interesting to inquire whether any land use innovations could be introduced at an early stage which would embody a preadaptive solution to the "tragedy of the commons" which so often accompanies the later stages of intensification. At the earlier stages the land in question is likely still to be under the stewardship of the local chief or kin group head. Perhaps this is the right person to approach with preadaptive land uses in mind.

Increase in yields per unit of land

When, in the course of swidden intensification, the fallow period becomes too short to allow adequate restoration of soil fertility and control of weeds, the swidden "degradation syndrome" sets in, driven by declining yields and the need to clear and plant larger fields to achieve a given production target, which in turn results in shorter fallows, further decline in yields, and so on. In the absence of opportunities to relieve population pressure by migration, the only way out of the degradation cycle is to adopt technological changes in farming practice which result in higher yields per unit of cultivated land. Such changes in technology have generally been accompanied in preindustrial (i.e. non-fossil fuel dependent) agriculture by declining returns to labor (Boserup 1965, 1981).

Thus, the development sequence depicted in Table 5 may be characterized as a "reluctant evolution," since farmers will quite rationally avoid the adoption of intensive technological innovations as long as extensive, less labor-requiring alternatives are still capable of satisfying their production objectives. Whether there are alternatives to arable crop intensification which offer higher returns to *both* land and labor is one of the central questions in agroforestry. Preliminary indications are quite promising (Raintree and Warner 1985).

In Stage 3, fallows continue to shorten but the length of the cropping period may remain the same or even lengthen through the use of crop rotations (Okigbo 1984). Field patterns have usually shifted by Stage 3 from the previous situation in which nearly all of the fields were fallowed to a mixed mosaic of permanently or semi-permanently cultivated fields on the best land, with recurrently cultivated fields on the less fertile land (Greenland 1974). Plow agriculture and manuring make their entry at this stage on the permanently or semi-permanently cultivated fields, as ways of increasing the productivity of cultivated land in the absence of adequate fallow. Farmers may also begin to exhibit a willingness to use commercial fertilizers, pesticides, weed controlling chemicals and other productivity increasing or labor saving inputs, *if* they are available and offer favorable returns on the investment and *if* cash savings or credit are available. The willingness and/or ability to adopt these technological innovations is often closely associated with cash cropping enterprises.

The differentiation of field types and cropping systems at varying levels of technology, the rise of cash cropping and the differential use of commercial inputs, along with the general trend of increasing population density toward more diversified economics based on specialization of production roles, all open up greater possibilities for economic stratification based on differential access to land of variable quality and/or differential adoption rates and opportunities for new cash crops or productivity increasing technologies. Such changes become more permanent when they are combined with supportive changes in land tenure.

One of the most significant ways in which increased socioeconomic disparity is expressed is in the differentiation of male and female production opportunities and responsibilities. (See Rocheleau, this volume.)

Stage 4, when all the land is in permanent cultivation, marks the end of shifting cultivation, although it is a natural outcome of swidden intensification. Fields are now used for at least one crop per year and with multicropping may be used continuously. In the absence of fallow to restore fertility and control weeds, industrial inputs such as fertilizer and herbicides, or alternatively, highly labor intensive practices are needed to sustain productivity. Under conditions of continued population increase, with sufficient industrial inputs to support very high levels of crop offtake it is problematic how much further the intensification of annual crop husbandry can be pushed.

Landlessness or near-landlessness is an inevitable result of unchecked population growth. Wage labor, land rent and tenancy become common, along with outmigration in search of alternative employment opportunities. If the rural areas are to be more than just subsistence havens for retired people, semi-dependent women and children and, indeed, if they are to rise to the challenge of supporting substantially higher expected rural population densities at adequate levels of maintenance, significant technological transformations of land use will have to occur. And these imply a host of far-reaching changes in tenure relations.

Agroforestry potentials and tenure issues

To clarify the concept and scope of agroforestry it may be useful to review the definition used by ICRAF.

> Agroforestry is a collective name for land use systems and technologies in which woody perennials (trees, shrubs, palms, bamboos, etc.) are deliberately combined on the same management unit with herbaceous crops and/or animals, either in some form of spatial arrangement or temporal sequence. In agroforestry systems there are both ecological and economic interactions among the different components (after Lundgren 1982).

From a project standpoint there are basically two ways of arriving at agroforestry: integrating trees into farming systems or farmers into forests.

Appropriately selected woody components may contribute to both the productivity and sustainability of farming systems on marginal land by enhancing organic matter production, maintaining soil fertility, reducing erosion, conserving water and creating a more favorable microclimate for associated crops and livestock. These "service roles" of trees in agricultural systems are above and beyond the direct "production roles" they might also play in supplying food, fodder, fuelwood, building materials, and a variety of raw materials for rural industry. Traditional land use practices also indicate the importance of agroforestry in maximizing and diversifying the productivity of even highly productive lands. Intensive agroforestry systems are most commonly found in areas with a long history of population pressure, indicating their general efficiency as a land use system. Whether on marginal or high potential lands, diversified agroforestry systems may be the most appropriate form of land use where land tenure constraints, lack of marketing infrastructure or an unfavourable political economy make it imperative for risk-reducing smallholders to try to satisfy most of their basic needs directly from the land resources under their control (Lundgren and Raintree 1983).

All tropical land use systems exhibit varying degrees of "leakiness" with respect to the cycling of nutrients held in the soil-vegetation complex (see Nair 1984), although such systems as irrigated rice paddies, permanent tree crops and forests are inherently more sustainable than others. It is a fundamental contention of agroforestry that trees have the potential to play many of the roles in tropical farming systems. The degree of spatial "infilling" can vary from slight (limited interstitial planting) to virtually complete (as in the home garden model). Essentially, the decision of how many and what kind of trees it is profitable to add to the existing pattern of land use depends on what useful niches for trees can be identified. An agroforestry "niche" in this sense has three components: 1) a functional *role* within the land use system, 2) a *place* within the landscape, and 3) a *time* within the cycle of land use or stagewise development of a particular system.

Although many of the recent research thrusts in agroforestry have been directed toward the integration of trees into farming systems, agroforestry also has a role to play in the preservation of forests and the improvement of forest management systems. By providing farmers with a means of producing fuelwood, timber, building poles and other "forest" products on farmland, agroforestry can significantly reduce the pressure demand on forests and natural woodlands. By doing this in ways that enhance and sustain agricultural productivity, agroforestry can also alleviate some of the pressure for the conversion of forest land into farmland. Moreover, the integration of farmers into forest management schemes through the use of "compromise" land use systems based on agroforestry may be one of the few realistic ways of sustaining forestry production on agriculturally pressured forest land (Raintree and Lundgren 1985).

Let us now consider some of the specific ways in which agroforestry might be used to address the problems and opportunities identified in the previous sections of this paper.

From shifting to permanent cultivation

Different agroforestry options and development pathways open up from different stages in the intensification sequence outlined in Table 5. Figure 2 summarizes what appear to be the most promising agroforestry approaches at different stages. Table 6 outlines in somewhat greater detail the suggested matching of agroforestry technologies to specific stages.

Table 6. Stages in the intensification of shifting cultivation and matching agroforestry options. Primary (P) and secondary (S) feasibilities indicated. Source: Raintree and Warner (1985).

	Stages of Intensification			
	1	**2**	**3**	**4**
Improved Fallows				
"Integral taungya"	P	S		
Economically enriched fallows	P	P		
Biologically enriched fallows		P	S	
Alley cropping/ mixed intercropping		P	P	
Tree Crop Alternatives				
Tree cash crops	P	P	S	S
Home gardens		S	P	P
Interstitial plantings		S	P	P

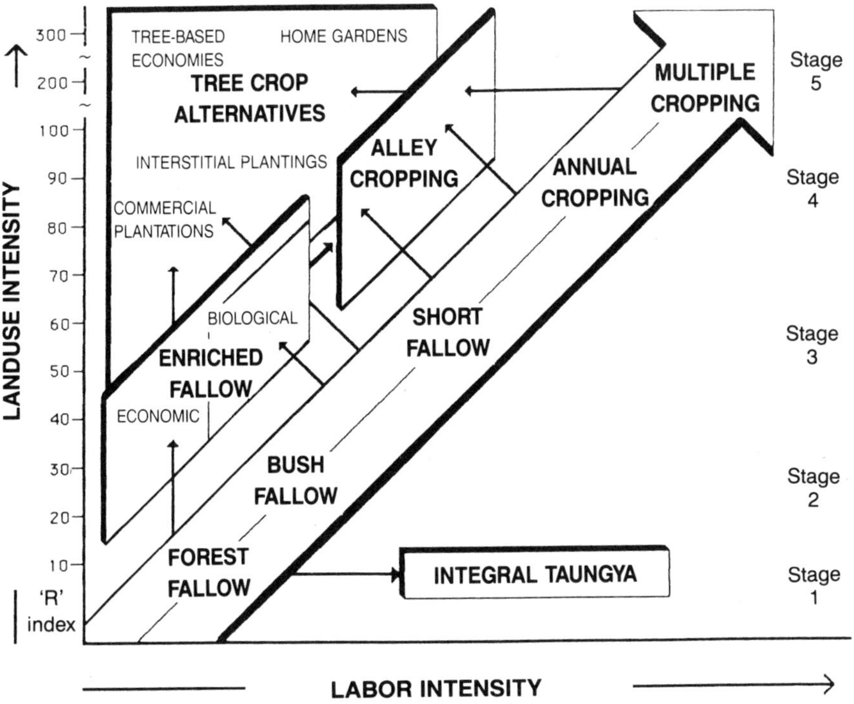

Figure 2. Agroforestry pathways for agricultural intensification in the tropics, with particular reference to the humid and subhumid zones. The R index (Joosten 1962) shown on the left gives an approximate indication of the land use intensity corresponding to the stages shown on the right. R = (C/C+F) x 100, where: C = cropping period, F = fallow period. The R index is also equivalent to the percentage of land in cultivation, as read from aerial photos. Boserup's (1981) treatment of R as a "frequency of cropping" index allows the interpretation to be extended to multicropping. For R | 100, R corresponds to the number of crops taken per year x 100. Source: Raintree and Warner (1985).

Integral taungya

In the classification of swidden systems, the distinction is made between "integral" and "partial" swidden. Partial swidden reflects "predominantly only the economic interests of its participants (as in some kinds of cash crop, resettlement, and squatter agriculture)," whereas integral systems "stem from a more traditional, year round, community-wide, largely self-contained, and ritually-sanctioned way of life" (Conklin 1957).

By analogy with integral shifting cultivation, the proposed concept of "integral taungya" is meant to invoke the idea of a land use practice which offers a more complete and culturally integrated approach to rural development: not merely the temporary use of a piece of land and a poverty level wage for labor, but a chance to participate equitably in a diversified and sustainable agroforestry economy.

The social aims of the proposed approach are high, and they are nowhere yet fully realized in practice; but perhaps the "Forest Village" schemes in Thailand come closest to the ideals of this concept (see Pragtong, this volume). In some variants of this approach, to make participation in the forestry effort more attractive to traditional shifting cultivators, the Thai foresters not only encourage the participants to grow long term perennial cash crops by widening the between-row spacing of the commercial forest species, but also allocate permanent agricultural plots to the farmers for use as they see fit. In addition, the scheme pays decent wages for a variety of work opportunities in the forestry sector of the village economy and provides a range of extension and community development inputs, such as housing assistance, clinics, schools, and places of worship (Boonkird et al. 1984). Far from an "exploitative" practice, this Thai variant of the taungya system shows promise of becoming a model example of what is meant by "integral taungya," although it is nowhere adequately documented in the literature. There is always the danger, of course, that the ideals of this approach could be subverted and that the banner of "integral taungya" could be used as a cover for politically motivated strategies of ethnic containment and "villagization."

Enriched fallows

The two variants of this approach, *economically enriched* fallows (which increase the economic utility of the fallow vegetation by enrichment with trees valued for cash or subsistence purposes) and *bio-*

logically enriched fallows (which enhance and accelerate the vegetative regeneration of soil fertility and control of weeds), become interesting to farmers somewhere around Stages 1-2 and 2-3, respectively, in the intensification sequence. Long fallow forest shifting cultivators are not likely to be interested in techniques to cope with soil fertility and weed control, since these are not yet likely to be a problem, although they may well respond to opportunities for economic benefits from improved fallows.

From the viewpoint of the shifting cultivator, the forest phase of the taungya cycle is equivalent to the fallow phase of the swidden cycle. To successfully marry the two types of production the phase lengths must match. Hence, the entry point of primary feasibility for integral taungya would seem to be in Stage 1 when fallow lengths are of the same order of magnitude as the growing period for commercial forest trees. However, it could possibly be introduced at Stage 2, providing that short-rotation (< 10 years) forest species are selected, or that agricultural practice can be technologically intensified to a degree which could accommodate the reduction in available fallow land that would accompany the planting of longer duration forest trees.

If fruit trees are planted instead of conventional forest trees, very long "fallow" durations may be required to sustain shifting cultivation, since farmers will be reluctant to cut down the fruit trees in the years of their heaviest bearing (which may extend from 10 to 100 years for some trees). Indeed, the planting of fruit or other valuable, long duration, continuously productive trees may be the shortest route to permanent removal of the land from the fallow cycle. This is not necessarily a bad thing, if the resulting land use mix (including biologically enriched short fallow or other means of meeting the production target of preferred field crops) is significantly more rewarding than the continuation of conventional swidden practice.

The basic feasibility of the economic enrichment approach is demonstrated by such indigenous examples as the fallow plantings of rattan by the Luangan Dayaks of Borneo (Weinstock 1983), of cedar and bamboo by the Lingnan Yao of China (Lebar et al. 1964), of casuarina by the Siane of New Guinea (Salisbury 1962), of gum arabic in the Sahel (Maydell 1980, and Seif el Din, this volume) and of multipurpose fallow "woodlots", whose species diversity exceeds that of the natural forest, by the Ifugao terrace builders of the Philippines (Conklin 1980). The latter agroforestry practice exemplifies not only a very rich, long duration fallow system but also the role which such systems can play in meeting requirements for ecologically sound upper watershed management. It should be noted, however, that the system seems to be breaking down in some localities because of accelerated cutting of wood for the tourist woodcarving industry, increased fuelwood pressure and a complex set of social changes associated with the loss of customary regulatory controls formerly exercised by the wealthier and more influential farmers whose influence is waning under the more equalitarian ethos of modern times (Eder 1982). In some parts of Ifugao, notably in the vicinity of the main tourist center, "disintensification" (Brookfield 1972) of traditional agriculture seems to be occuring as a result of increased opportunities for non-farm employment and other modernizing influences.

Biologically enriched fallow practices using *Acioa barteri*, *Anthonotha macrophylla*, *Alchornea cordifolia*, *Gliricidia sepium* and *Leucaena leucocephala* have been reported by Benneh (1972), Okigbo and Lal (1979), Getahun et al. (1982), Agboola et al. (1982) and Dijkman (1950). For further discussion and examples of fallow enrichment practices, see also Olofson (1983), FAO (1984), and Raintree and Warner (1985).

The most obvious implication of these practices for tenure is the increased premium their adoption would place on rights of exclusive harvest and reuse by the management unit which undertakes the labor investment of the fallow improvement. Without secure rights of this type, it is unlikely that the benefits would be worth the effort. The magnitude of the potential benefits of the approach would seem to justify efforts to bring about supportive tenure changes, with the caveat always in mind, however, that the system could be abused by individuals as a strategy for grabbing land.

Alley cropping and other intercropping systems

If the preceding discussion stretches the conventional concept of "fallow" beyond its normal usage, the extension of the approach into the "continuous fallow" processes (G.F. Wilson personal communication) of "alley cropping" goes even further toward a thoroughly functional reinterpretation of the concept for tropical conditions. The beneficial effects of soil restoring trees on agricultural crops can be effected by associating the two components in *time*, as in the sequential practice of fallow rotation, or in *space*, through simultaneous association of trees and field crops.

Alley cropping may be defined as a "zonal" approach (Huxley 1980, Huxley and Raintree 1983) to agroforestry, in which field crops are planted in the alleys between hedgerows of nutrient-cycling trees or shrubs, which are kept pruned throughout the cropping season to control shading and below ground competition while providing green manure and mulch material for the benefit of the associated crops. Fodder and fuelwood might be taken as by-products of the system, but the basic aim is to fulfill a "service function" (Torres 1983) within the arable farming system. The term "alley cropping" was coined by researchers at the International Institute of Tropical Agriculture in Ibadan (Wilson and Kang 1980), but the technique itself appears to have originated in indigenous practice some five decades earlier on the island of Timor under the direction of the Raja of Amarasi (Metzner 1981, Olofson 1983). The economic benefits of various experimental alley cropping systems have been examined by Raintree and Turay (1980), Verinumbe et al. (1984), Hoekstra (1983) and Ngambeki and Wilson (1984). It has been described in a recent FAO publication as "possibly the most versatile, effective, and widely adoptable of recent innovations in conservation farming" (FAO 1984).

More intimate "mixed" arrangements of trees and crops are also found in traditional practice, the outstanding example being the association of *Acacia albida* with dryland grain crops in the Sahel (Felker 1978, Weber and Hoskins 1983, NAS 1983), where the yields of crops grown in proximity to the tree are typically double those of crops grown outside the tree's sphere of influence. Nitrogen fixation, micro-climatic benefits and the peculiar "reverse phenology" of the tree (which leafs out in the dry season and drops its leaves at the beginning of the rainy season, thus nicely accommodating itself to the requirements of crop cultivation), account for part of the yield increase. But an important tenure-related effect is the concentration of manure in the vicinity of the trees as a result of the gathering of livestock under thin shade during the dry season to consume the nutritious pods produced at this time. Presumably, the benefits to the arable crops would be reduced if traditional grazing rights were restricted.

Since the deliberate establishment of new alley cropping or mixed intercropping systems may represent a substantial investment of labor and other resources, security of tenure becomes an important precondition of adoptability. This does not necessarily mean that the benefits of multipurpose trees (see FAO 1984:32 for selection criteria) could not conceivably be shared by claimants with different land and tree utilization rights (see Fortmann this volume). But it is obviously the case that the incentives for adoption of these more-or-less permanent improvements would be more effective if exclusive use rights were vested with the innovating management unit; although, as discussed below, it might be advantageous to allow controlled grazing by livestock (owned by pastoralists or other farmers) during the dry season. Insofar as the planting of trees establishes a legal claim to the land on which they are planted, the management unit in question will, in most cases, also have to be the land holding unit. This latter point was brought home to me while conducting farm trials of alley cropping systems in Nigeria, where some of the participants had to drop out when the kin group steward of the land "borrowed" for the trials learned that trees were to be planted and withdrew the usufruct rights.

For these reasons, and for reasons associated with the relatively higher labor requirements of the practice (as compared to planted fallows), intensive alley cropping systems are not likely to become very attractive to farmers until the short fallow or permanent cultivation stages (3 and 4) of the intensification sequence, when ecological demands and supportive tenure adjustments make it necessary and possible. Again, providing the system is not abused as a way of grabbing excessive amounts of land, supportive tenure adjustments would seem justified.

One way of effecting a smooth adjustment of agroecological and tenure factors associated with alley cropping would be to take a *phased approach* to the adoption of the system, based on the concept of an "optional pathway of intensification" (Raintree 1980, 1983b, FAO 1984, Raintree and Warner 1985). Starting with a fallow enrichment approach at Stage 2, tree species could be introduced which have both economic and biological fallow improving properties. By planting the selected trees in hedgerows at appropriate between-row spacing (which could be adjusted for effective erosion control on sloping lands), the way would be clear for an intensification of the fallow practice into semi-permanent or permanent alley cropping at Stages 3 and 4. As a final measure of intensification, undertaken under conditions of very high population pressure by the children or grandchildren of the original shifting cultivators, the installed "green manure factories" could be maintained in place and a variety of economically valuable upperstorey trees could be added to the system. In this last phase of intensification the system might come to resemble the architectural complexity and economic efficiency of the multistorey home garden, so often found in densely settled areas of the tropics.

If the motivation exists, there is no inherent reason why the scheme of intensification envisaged here could not be run "ahead of itself" to generate higher incomes for industrious rural families well in advance of population-pressured necessity. Of course, this may result in an accelerated emergence of income disparities, but this would be true of virtually any productivity enhancing innovation and should not be used as an argument to squelch the innovation. Projects could take measures to insure equality in the distribution of opportunity to adopt the system (which, in itself, may involve land reforms).

Tree crop alternatives

Even with the aforementioned improvements as an efficiency enhancing adjunct to the use of fertilizers and other modern agricultural inputs, there is a limit to what can reasonably be expected from the intensification of tropical field crop systems.

Several writers have argued the superiority of tree-based cropping systems over arable cropping systems for difficult farming situations (Smith 1950, Douglas and Hart 1976, Felker and Bandurski 1979, Mollison and Holmgren 1981, Bowers 1982, Chambers 1984), but nowhere is the argument more compelling than in humid tropical environments where the temperate cropping system model has established an unjustified hegemony over the imagination of agriculturalists and land use planners.

> The really critical problems to be overcome, then, if high productivity resources exploitation systems are to be developed for these life zones, would seem to be those of achieving an efficient economic utilization of the diversity of massive, fast-growing, perennial plant species and vegetational types which these environments are themselves efficient at producing (Tosi and Voertman 1964).

The viability of production systems in which trees and other perennials, rather than annual field crops, are the main focus of the land use economy are indicated by the widespread practice of "home gardens" in areas of high population density, where conditions are such as to favor the most efficient land use systems. In his study of Ibo farming practices in Nigeria, where population densities may reach as high as $1000/km^2$, Lagemann (1977) found a strong correlation between population density and the importance of multistorey compound gardens. This is largely explained by the fact, according to

Lagemann's figures, that the output from compound gardens is 5 to 10 times greater in monetary terms than from the outfield plots, and the returns to labor are 4 to 8 times higher. In Java, where various tree garden types exist in a mosaic with wet rice paddies and rainfed arable crops (Penny and Singarimbun 1973, Wiersum 1982, Hunink and Stoffers 1984), home gardens may provide more than 20 percent of household income and 40 percent of household caloric requirements (Stoler 1978).

Highly integrated tree-based economies have been reported from Indonesia, involving the near total exploitation of the lontar palm (*Borassus sudaicus*) on Roti and Savu islands (Fox 1977) and the domestication of *Shorea javanica* in southern Sumatra (Torquebiau 1984), a species which silviculturists have regarded as too difficult for plantation purposes. The role of trees in these economies is exceeded, perhaps, only by the tree based systems of the Pacific atolls and low islands (Barrau 1971, Schirmer 1983), where even the drinking water may be provided by a tree (i.e. the coconut palm). In this connection, the underexploited potential of the palms is so striking as to deserve special mention (see Johnson 1983).

Horticultural tree crops for cash purposes are extensively planted by smallholders as an outgrowth of shifting cultivation in many parts of the world, notably the oil palm, cacao, coffee and cola nut plantations of West Africa, covering as much as 67 percent of the land in southern Nigeria (Getahun et al. 1982), and the coconut, rubber, oil palm, cacao and coffee plantations of smallholders in Southeast Asia (Pelzer 1978, Nair 1979, Liyanage et al. 1984, Dove 1983).

The transition to tree crop based systems is not equally feasible from all stages in the main sequence of intensification in tropical land use. There are few ecological constraints in Stage 1 and 2 to the planting of extensive areas to tree crops, although the economic incentives will generally have to be rather attractive, since leisure for social activities is likely to be highly valued in integral swidden societies at this stage of development. Nevertheless, as Dove (1983) has pointed out, extensive cash cropping of trees is a common feature of many relatively long fallow swidden systems. At Stages 3 and 4, however, the transition to tree crops is less easily achieved due to the commitment of land to other uses and the relatively long lag time between planting and first harvest. Nevertheless, the economic motivation for a transition to tree crop systems may be quite high and "taungya" practices can ease the burden of the establishment phase by providing early returns of interplanted field crops.

Since the planting of trees as cash crops will often take land out of food crop production, after Stage 1 extensive plantings must usually be accompanied by some form of field crop intensification. On the whole, however, tree crops may offer higher returns to a more balanced mix of inputs than annual field crops (as suggested by Figure 3).

Needless to say, the investments required in tree planting require secure and fairly exclusive tenure by the planters over both the trees and the land on which they are planted, although it is again conceivable that various usufruct and harvesting rights could be subdivided among different claimants (see Fortmann, this volume). For example, grazing rights could be exercised by livestock keepers, with or without payment to the land holder, once the critical tree establishment phase is passed.

Interstitial tree planting

It has been repeatedly suggested throughout this paper that tree planting need not always be in competition with field crops for land. The planting of fertility-enhancing trees is one case in which the crops may actually benefit from association with the trees. Examples of such "complementary" economic relationships abound in practice, but they remain underdocumented and little understood, and planners continue to speak of the assumed competition between trees and food crops as if it were, alas, an inescapable fact of life. Examples of "supplementary" economic relationships, in which the trees and other crops have negligible or neutral interaction, can also be cited. Even when there is a "competitive" relationship between trees and other crops, a certain amount of controlled mixing may be justified in terms of the net economic yield of land use systems oriented toward the production of a diversity of products (Arnold 1983, Raintree 1983a, Hoekstra 1985).

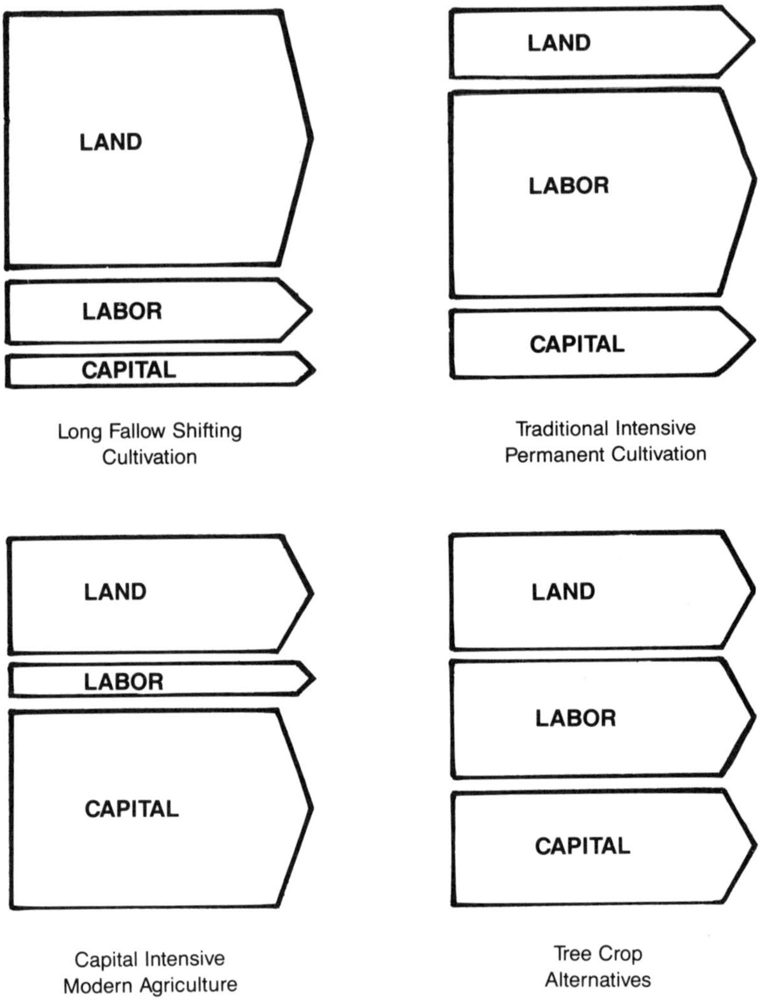

Figure 3. Relative factor proportions of different tropical land use systems. One of the potential advantages of trees as a productive form of capital for the self-reliant smallholder is their partial substitution for labor and less accessible capital inputs thus making it possible to aspire to a more balanced mix of production factors.

The planting of trees in interstitial locations within farms, along farm boundaries and internal borders, or along roadsides, watercourses, and on wasted or underutilized lands in the general landscape, offers a special opportunity for supplementary production, since plantings at these locations (almost regardless of the biological competitiveness of the trees) may often be undertaken with little or no opportunity cost to other land uses. For example, a recent aerial photo analysis of a watershed in a fairly densely settled farming community in the subhumid midlands of Kenya indicated that, if existing linear features of the landscape (pathways, watercourses, farm boundaries and internal borders) were fully utilized for planting of appropriate trees and shrubs, some 50 percent of the fuelwood and 40 percent of the fodder requirements of the households in the area could be supplied by these hedgerows alone, with very little competition with existing agricultural land uses (Rocheleau and Hoek 1984).

Can it be assumed that the tenure issues arising in connection with interstitial plantings will be as benign or readily solvable as the technological problems? Possibly not, since in the case of boundary plantings the *boundaries* per se might be the source of dispute, and trees planted on degraded or "underutilized" lands in the general landscape may arouse the concerns of those currently enjoying gathering or grazing rights in the commons. Perhaps some kind of common property approach to multiple use of these lands may be feasible. The opportunities for the satisfaction of diverse interests might be enhanced in such cases by the use of appropriately selected multipurpose trees. One social organizational approach which might be worth exploring is the partitioning of planting responsibilities and exclusive harvest rights among individual members of an interhousehold working group, organized on a neighborhood basis and operating at a larger-than-farm scale (Rocheleau 1984).

As regards boundary disputes per se, it is an important consideration that trees may take on legal meaning as boundary markers. In Kakamega District in Western Kenya, for example, it doesn't matter where you might locate your barbed wire fence, the boundaries of the farm are always judged by the location of the obligatory *Euphorbia tirucali* hedge. Even where boundaries are well established, boundary planting of trees may lead to problems with the neighbors. There are, perhaps, two approaches to the resolution or avoidance of such conflicts. One is to plant only valuable fruit or fodder trees and allow the neighbors to take the share of the produce which extends into or falls on their property. The other approach is to choose trees which are as neutral as possible, offering little shade or other competition to whatever might be on the other side of the boundary. The disadvantage of the second approach is that it may be taken to the point that only useless trees can be planted, as in the case of the comparatively useless *Euphorbia tirucali* hedges of Kakamega (apologists will argue that the tree *can* be used as an emergency fuel, but there are many better fuelwood species to choose from). Nevertheless, it seems to be the very neutrality, indeed, the comparative uselessness of this tree which gives it its unique legal significance as a boundary marker in Kakamega District.

Where land use patterns and tenure rules are undergoing changes in adaptation to population pressures or other factors, the boundary marking role of trees can have either positive or negative social effects, depending on who is planting them to establish what kind of claims to land and whether or not such claims are considered legitimate. On the one hand, trees can be used to consolidate tenurial aspects of ecologically necessary and beneficial changes, while on the other hand they might be used for out and out land grabbing.

The planting of trees at interstitial locations within farms might seem to be a wholly positive development, since this is one relatively painless way to increase the supply of tree products for household consumption, sale or savings on land which is under the direct control of the household. Unfortunately, this is where many of the gender-related tenure problems arise. It all depends on what *kind* of trees are planted and where. In the Central Highlands of Kenya, there are men's trees (timber, cash crop) and women's trees (fuelwood, fodder, subsistence). In Kakamega all trees are owned by the men, and there are strong cultural prohibitions against the planting and felling of trees by women. It is said that if a woman plants a tree she will become barren and her husband will die (Chavangi 1984). Of course, women manage to get around these restrictions, by various ruses, and the wisdom of these cultural rules is beginning to be openly questioned. But the Kakamega case is illustrative of how it is possible to have a landscape full of trees (mostly eucalyptus and cypress) and yet the women must rely on leaves and twigs for much of the domestic energy supply.

Agropastoral interactions

The ecological problems posed by the dry regions of the tropics make the integration of trees into land use systems especially imperative and, at the same time, rather more difficult to achieve than in the wetter zones. Two factors are primarily responsible for the difficulties: 1) aridity itself, which increases the risks and the costs associated with successful tree establishment (usually requiring the laborious planting of individual seedlings; whereas, in the humid tropics direct seeding often suffices), and 2) browsing damage by livestock. The investment requirements associated with the former imply the need for security of tenure over the trees, and the latter implies land use conflict over customary grazing rights.

Providing that ways can be found to solve the tenure problems, the integration of trees into land use systems in the dry zones offers a number of possiblities for improving the linkage between agricultural and pastoral elements of the economies of these areas, both within and between management units. The trees themselves, properly selected and managed for multiple benefits, may provide the means of resolving or at least mitigating some of the most prominent tenure conflicts.

The role of trees in mixed farming systems

The expansion of farming into fragile dryland environments poses special problems for crop husbandry: conservation and efficient use of limited soil moisture, maintenance of soil structure and fertility, control of wind and water erosion, and provision of feed for draft animals. One of the keys to successfully coping with the first three of these problems is the maintenance of adequate soil organic matter and nitrogen (a recent analysis of the role of nitrogen in plant water use efficiency by Felker et al. 1980 has suggested that nitrogen may be more limiting than rainfall in many arid areas). Conventional methods of green manuring and mulch farming have generally been limited by the difficulty of growing enough herbaceous biomass for return to the soil and by the competition with food crops for water, land and labor. Likewise, the usefulness of herbaceous fodder crops as a source of supplemental feed to bridge the dry season fodder gap is limited by their low protein quality at this time of year and, again, by the competition for water, land and labor with food crops.

Multipurpose trees, particularly the nitrogen fixing species so well adapted to dryland conditions, offer several advantages over herbaceous sources of organic matter, nitrogen and fodder: 1) they are generally more drought tolerant than herbaceous plants, 2) they have higher feed value during the dry season (particularly the pod producing species) and are thus better able to ensure the strength of draft animals at the beginning of the rains, 3) they can be grown at interstitial locations on the farm or in association with crops without replacing them, 4) appropriately selected and arranged with respect to the crops, they can offer microclimatic benefits associated with windbreak effects and reduced evaporation, 5) they can produce food, fuelwood, building materials, and other directly useful by-products while performing their service roles on the farm, and 6) as a form of standing capital, they can serve as a source of convertible savings for emergency needs, including "famine foods."

In this latter role, trees can be a partial substitute for livestock, whose main role in Africa is "savings on the hoof." But they can also strengthen the role of livestock on the farm by enhancing the fodder-manure linkage and, if used as living fences, by providing an affordable means of reducing crop damage by uncontrolled grazing. These benefits can also be applied to livestock owned by others. In the case of interactions between farmers and pastoralists, the growing of additional, high quality dry season fodder on the farm and the use of living fences to control livestock access could potentially go a long way toward relieving the main sources of tenure conflict between the two land use systems. The main constraint to the realization of these benefits is, of course, the need to restrict livestock access, often by social means, during the establishment phase of the trees.

Thus, although we may be able to envisage a partial "technological fix" for certain agro-pastoral tenure problems, the path to these ends must be cleared by social means.

The role of trees in pastoral systems

While trees may have a role in bringing about an accommodation between pastoralists and farmers, pastoralists are not well advised to wait for farmers to come to their rescue by planting trees. Although it will usually require a much greater feat of "social engineering" (Cernea 1985) to promote effective tree planting by pastoralists, there are a number of pastoral situations in which the need for trees would seem to loom so large as to justify cautious optimism for the success of well planned projects.

One of these situations is overgrazing a- round dry season water sources. The Ferlo Project in Senegal (Maydell 1980, NAS 1983) and the "Land Management Near Wells" project in Niger (Weber and Hoskins 1983) have been exploring various technical and social models for mitigating the ecological effects of herd concentrations around boreholes.

Another common situation is overgrazing around pastoral home areas and the problems associated with the provision of feed to young and sick animals that are kept in these areas. The Maasai "olopololi" is a dry season grazing reserve maintained jointly by several households for such animals in the vicinity of their residential bomas. Since this area is protected from grazing during the rainy season, it might be the right niche for the planting of supplemental fodder trees. Some Maasai households in the Group Ranches are beginning to experiment with crop production in small, thorn-fenced gardens near the bomas. This also would be a natural site for the growing of supplemental fodder trees to meet dry season feed requirements (Nambombe 1984, Mhungu 1984). It would be especially useful to use such fodder to feed smallstock and weaning calves, since the expected production would be roughly commensurate with the volume requirements of the smaller herd animals and, in the latter case, the milk savings effected by substitution of tree fodder would directly benefit the household, which is normally in direct competition with the calves for milk (B.E. Grandin personal communication).

These are the kinds of modest agroforestry proposals that it might be worth exploring with pastoralists and, insofar as they are oriented toward the development of small areas under control of one or a small group of households, would not seem to pose serious tenure difficulties. The challenge of more general range improvements for the benefit of the mature herd animals presents much more serious problems, insofar as it would involve common grazing lands. However, to cite the Maasai example again, such pod producing legumes as *Acacia tortilis* occur naturally on the open range in Maasailand and in some cases their exploitation is controlled by the local group.

According to the accounts given by Maasai informants to an ICRAF research trainee (v. Nambombe personal communication) the "courtesy" rights accorded by one group ranch to members of a neighboring ranch to utilize the nutritious *Acacia tortilis* pods underwent an interesting change during the recent drought. Normally they were allowed to shake the tree, thus causing more pods to fall to the ground for their animals, but as the drought became more serious the right to the pods was restricted to those which fell to the ground naturally. The right to shake the tree to make more pods drop was reserved to the local group ranch, if, in fact, the Maasai are exercising this kind of regulatory control over the use of the pods of such trees, might they not then be encouraged to undertake steps toward the domestication and artificial propagation of this valued range resource?

But who would undertake such tree planting for the general benefit? As W.R. Bentley (personal communication) has observed, the "tragedy of the commons" in India is not so much a problem of overexploitation as one of underinvestment. Posing the problem in this way suggests that one approach to encourage greater investment in the commons might be to identify the least costly method of investment. Now, it is a fact that such valuable rangeland trees as *Acacia albida*, *Prosopis spp.* and other pod producing legumes can be, and often are, propagated by livestock (the predominant tree in the arid rangelands of Rajastan is *Prosopis cineraria*, an introduced exotic which spread naturally, quite without the benefit of tree planting projects). What is to prevent a rangeland management project from supplying such pods in quantity to pastoralists to feed to their herds and see what happens? No doubt the establishment percentage would be low, but so would the herder's investment, and the result, probably long after the project had come and gone, might be quite significant. We don't know whether this would work because we don't know of any project that has tried it. The point is that there might be some very simple technological fixes to some of the problems of the commons.

At the other end of the scale is quite another kind of intervention in pastoral rangelands. Interest in dryland biomass energy plantations has quickened in recent years (Foster and Karpiscak 1983, Felker 1984). From the pastoralist's point of view, the problem with such schemes is that they tend to represent yet another assault on traditional tenure rights, but is it really inconceivable that pastoral populations would participate in integrated plantation schemes? The problem of identifying a basis of shared interest would seem, from an agroforestry point of view, to reduce to the problem of identifying the *right multipurpose trees*. Why not choose an energy tree from among several outstanding desert biomass producers that happen also to produce copious quantities of high quality dry season pods. Again, *Prosopis* would seem to exemplify the appropriate ideotype (Felker and Bandurski 1978, Felger 1977). The reasons for choosing a pod or other fruit producing fodder species is that the utilization of this by-product need not reduce the eventual woody biomass harvest of the plantation, that it could make a big difference in the dry season carrying capacity of the range, and that such a choice might make all the difference in the world to the survival rate of the trees in pastoral areas. In this connection it is interesting to note that present day energy experts are coming to the conclusion that for biomass energy converters to realize their full economic potential, they should be conceived in terms of by-product and co-conversion schemes (Williams 1985, Reddy 1985). Does this principle not also apply to the production technology?

Could not industrial agroforestry schemes along these lines provide a simple, equitable, all round solution to the related developing country problems of biomass energy supply, decentralization of rural industry and participation of pastoralists in national development?

Perhaps it is well to end here on an imaginative note and with a question, since the purpose of this paper has been to provide some images to serve as a background to the consideration of tenure questions in agroforestry.

Notes

1. The account of the evolution of tropical agricultural systems from shifting cultivation to permanent cropping is based largely on Raintree and Warner (1985).

References

Agboola, A.A., G.F. Wilson, A. Getahun and C.F. Yamoah. 1982. Gliricidia sepium: a possible means to sustained cropping. *In* L.H. MacDonald (ed). *Agro-forestry in the African Humid Tropics.* United Nations University. Tokyo.

Allan, W. 1965. *The African Husbandman.* Oliver and Boyd. Edingburgh.

Arnold, J.E.M. 1983. Economic considerations in agroforestry projects. *Agroforestry Systems* 1(4):299-312.

Barrau, J. 1971. *Subsistence Agriculture in Polynesia and Micronesia.* Bishop Museum. Honolulu.

Benneh, G. 1972. Systems of agriculture in tropical Africa. *Economic Geography* 48(3):244-257.

Boonkird, S.A., E.C.M. Fernandes and P.K.R. Nair. 1984. Forest villages: an agroforestry approach to rehabilitating forest land degraded by shifting cultivation in Thailand. *Agroforestry Systems* 2(2):87-102.

Boserup, E. 1965. *The Conditions of Agricultural Growth.* Aldine. Chicago.

Boserup, E. 1981. *Population and Technology.* Basil Blackwell. Oxford.

Bowers, R.D. 1982. Agricultural tree crops as a no-tillage system. In L.H. MacDonald (ed). *Agro-forestry in the African Humid Tropics.* United Nations University. Tokyo.

Brookfield, H.C. 1972. Intensification and disintensification in Pacific agriculture: a theoretical approach. *Pacific Viewpoint* 15:30-48.

Cernea, M.M. 1985. Land tenure and the social units sustaining alternative forestry development strategies. In M.M. Cernea (ed). *Putting People First: Sociological Variables in Development Projects.* Oxford University Press. Oxford.

Chambers, R. 1984. To the hands of the poor: water, trees and land. Discussion Paper No. 14. Ford Foundation. New Delhi.

Chavangi, N. 1984. Cultural aspects of fuelwood procurement in Kakamega District, Kenya. Working Paper No. 4. Kenya Fuelwood Development Programme/Beijer Institute. Nairobi.

Coale, A.J. 1974. The history of the human population. In Scientific American (ed). *The Human Population.* W.H. Freeman and Company. San Francisco.

Conklin, H.C. 1957. *Hanunoo Agriculture.* FAO Forestry Development Paper No. 12. FAO. Rome.

Conklin, H.C. 1980. *Ethnographic Atlas of Ifugao: A Study of Environment, Culture and Society in Northern Luzon.* Yale University Press. New Haven and London.

Dijkman, M.J. 1950. Leucaena--a promising soil erosion control plant. *Economic Botany* 4:337-347

Douglas, J.S. and R.A. de J. Hart. 1976. *Forest Farming*. Watkins. London.

Dove, M.R. 1983. Theories of swidden agriculture and the political economy of ignorance. *Agroforestry Systems* 1(2):85-100.

Eder, J.E. 1982. No water in the terraces: agricultural stagnation and social change at Banaue, Ifugao. *Philippine Quarterly of Culture and Society* 10:101-116.

Enabor, E.E. 1974. Socio-economic aspects of taungya in relation to traditional shifting cultivation in tropical developing countries. In *Shifting Cultivation and Soil Conservation in Africa*. FAO Soils Bulletin No. 24. FAO. Rome.

FAO. 1981. *Agriculture: Toward 2000*. FAO. Rome.

FAO. 1984. *Changes in Shifting Cultivation in Africa*. FAO Forestry Paper No. 50. FAO. Rome.

Felger, R.S. 1977. Mesquite in Indian cultures of southwestern North America. In B.B. Simpson (ed). *Mesquite: It's Biology in Two Desert Scrub Ecosystems*. Dowden, Hutchinson and Ross. Stroudsberg, Pennsylvania.

Felker, P. 1978. *State of the Art: Acacia albida as a Complementary Permanent Intercrop with Annual Crops*. U.S. Agency for International Development. Washington.

Felker, P. 1984. Economic, environmental and social advantages of intensity managed short rotation mesquite (*Propopis spp.*) biomass energy farms. *Biomass* 5(1):65-77.

Felker, P. and R.J. Bandurski. 1979. Uses and potential uses of leguminous trees for minimal energy input agriculture. *Economic Botany* 33: 172-184.

Felker, P., P.R. Clark, J. Osborn and G.H. Cannel. 1980. Nitrogen cycling-water use efficiency interactions in semi-arid ecosystems in relation to management of tree legumes (*Prosopis*). *International Symposium on Browse in Africa*. ILCA. Addis Ababa.

Foley, G. and G. Barnard. 1984. *Farm and Community Forestry*. Earthscan Energy Information Programme Technical Report No. 3. International Institute for Environment and Development. London.

Fortmann, L. and J. Riddell. 1985. *Trees and Tenure: An Annotated Bibliography for Agroforesters and Others*. Land Tenure Center and ICRAF. Madison and Nairobi.

Foster, K.E. and M.M. Karpiscak. 1983. Arid land plants for fuel. *Biomass* 3(4):269-285.

Fox, J.J. 1977. *Harvest of the Palm: Ecological Change in Eastern Indonesia*. Harvard University Press. Cambridge, London.

Getahun, A., G.F. Wilson and B.T. Kang. 1982. The role of trees in farming systems in the humid tropics. In L.H. MacDonald (ed). *Agro-forestry in the African Humid Tropics*. United Nations University. Tokyo.

Gourou, P. 1966. *The Tropical World*. 4th edition. Longmans. London.

Greenland, D.J. 1974. Evolution and development of different types of shifting cultivation. In *Shifting Cultivation and Soil Conservation in Africa*. FAO Soils Bulletin No. 24. FAO. Rome.

Harrison, P. 1984. A new framework for the food security equation. *Ceres* 17(2):25-36.

Hoekstra, D.A. 1983. An economic analysis of a simulated alley cropping system for semi arid conditions, using micro computers. *Agroforestry Systems* 1(4):335-346.

Hoekstra, D.A. 1985. The use of economics in diagnosis and design of agroforestry systems. Working Paper No. 29. ICRAF. Nairobi.

Hunink, R.B.M. and J.W. Stoffers. 1984. *Mixed and Forest Garders on Central Java: An Analysis of Socio-Economic Factors Influencing the Choice Between Different Types of Landuse*. Department of Geography of Developing Countries. Utrecht University. Utrecht.

Huxley, P.A. 1980. Agroforestry research: the value of existing data appraisal. In T. Chandler and D. Spurgeon (eds). *International Cooperation in Agroforestry*. ICRAF. Nairobi.

Huxley, P.A. and J.B. Raintree. 1983. Plant arrangement considerations. In resources for agroforestry diagnosis and design. Working Paper No. 7. ICRAF. Nairobi.

Johnson, D.V. 1983. Multi-purpose palms in agroforestry: A classification and assessment. *International Tree Crops Journal* 2(3/4):217-244.

Joosten, J.H.L. 1962. Wirtchaftlich und agrarpolitische aspekte tropischer landbausysteme. Institut fur Landwirtschaftliche Betriebslehre. Gottingen.

Kio, P.R.O. 1972. Shifting cultivation and multiple use of forest land in Nigeria. *Commonwealth Forestry Review* 51(2): 144-148.

Lagemann, J. 1977. *Traditional African Farming Systems in Eastern Nigeria*. Weltforum Verlag. Munchen.

Lebar, F.M., G.C. Hickey and J.K. Musgrave (eds). 1964. *Ethnic Groups of Mainland Southeast Asia*. Human Relations Area Files Press. New Haven.

Liyanage, M. de S., K.G. Tejwani and P.K.R. Nair. 1984. Intercropping under coconuts in Sri Lanka. *Agroforestry Systems* 2(3):215-228.

Lundgren, B. 1982. Introduction. *Agroforestry Systems* 1(1): 3-6.

Lundgren, B. and J.B. Raintree. 1983. Sustained agroforestry. In B. Nestel (ed). *Agricultural Research for Development: Potentials and Challenges in Asia*. ISNAR. The Hague.

Maydell, H.J. von. 1980. The development of agroforestry in the Sahelian zone of Africa. In T. Chandler and D. Spurgeon (eds). *International Cooperation in Agroforestry*. ICRAF. Nairobi.

McGown, R.L., G. Haaland and C. de Haan. 1983. The interaction between cultivation and livestock production in semi-arid Africa. In A.E. Hall, G.H. Cannell and H.W. Lawton (eds). *Agriculture in Semi-Aric Environments*. Springer-Verlag. Berlin.

Metzner, J. 1981. Innovations in agriculture incorporating traditional production methods: the case of Amarasi (Timor). *Applied Geography and Development* 17:91-107.

Mhungu, J.A. 1984. Land use and production activities in relation to cattle feed and calf/people competition on a Maasai group ranch. On-the-Job Training Programme Final Seminar Report. ICRAF. Nairobi.

Mollison, B. and D. Holmgren. 1981. *Permaculture One: A Perennial Agriculture for Human Settlements*. International Tree Crops Institute. Winters, California.

Nair, P.K.R. 1979. *Intensive Multiple Cropping with Coconuts in India*. Verlag Paul Parey. Berlin and Hamburg.

Nair, P.K.R. 1984. *Soil Productivity Aspects of Agroforestry*. Science and Practice of Agroforestry No. 1. ICRAF. Nairobi.

Nair, P.K.R. 1985. Classification of agroforestry systems. Working Paper No. 28. ICRAF. Nairobi.

Nair, P.K.R. (ed). 1987. *Agroforestry Systems in the Tropics*. Martinus Nijhoff. The Hague.

Nambombe, V. 1984. Diagnosis of agroforestry potentials in a pastoral system: Kiboko Group Ranch, Kajiado District, Kenya. On-the-Job Training Programme Final Seminar Report. ICRAF. Nairobi.

NAS, 1983. *Agroforestry in the West African Sahel*. National Academy of Sciences. Washington.

Ngambeki, D.S. and G.F. Wilson. 1984. Economic and on-farm evaluation of alley cropping with Leucaena leucocephala. 1980-1983 Activity Consolidated Report. Farming Systems Program. IITA. Ibadan.

Okigbo, B.N. 1984. Cropping systems and rotations development for improving shifting cultivation and related intermittent production systems in tropical Africa. In *Improved Production Systems as an Alternative to Shifting Cultivation*. FAO Soils Bulletin No. 53. FAO. Rome.

Okigbo, B.N. and R. Lal. 1979. Soil fertility maintenance and conservation for improved agroforestry systems of the lowland humid tropics. In H.O. Mongi and P.A. Huxley (eds). *Soils Research in Agroforestry*. ICRAF. Nairobi.

Olofson, H. 1983. Indigenous agroforestry systems. *Philippine Quarterly of Culture and Society* 11:149-174.

Pelzer, K.J. 1978. Swidden cultivation in Southeast Asia: historical, ecological, and economic perspective. In P. Kunstadter, E.C. Chapman and S. Babhasri (eds). *Farmers in the Forest: Economic Development and Marginal Agriculture in Northern Thailand*. University of Hawaii Press. Honolulu.

Penny, D.H. and M. Singarimbun. 1973. *Population and Poverty in Rural Java: Some Economic Arithmetic from Sriharjo*. Cornell International Agricultural Development Monograph No. 41. Cornell University. Department of Economics. Ithaca, New York.

Raintree, J.B. 1980. Conservation farming with multipurpose tree legumes: an underdeveloped branch of tropical agroforestry research. In L. Buck (ed). *Proceedings of the Kenya National Seminar on Agroforestry*. ICRAF. Nairobi.

Raintree, J.B. 1983a. Bioeconomic considerations in the design of agroforestry cropping systems. In P.A. Huxley (ed). *Plant Research and Agroforestry*. ICRAF. Nairobi.

Raintree, J.B. 1983b. The concept of an optimal pathway of intensification. In Resources for agroforestry diagnosis and design. Working Paper No. 7. ICRAF. Nairobi.

Raintree, J.B. 1984. A diagnostic approach to agroforestry design. In K.F. Wiersum (ed). *Strategies and Designs for Afforestation, Reforestation and Tree Planting*. PUDOC. Wageningen.

Raintree, J.B. and B. Lundgren. 1985. Agroforestry potentials for biomass production in integrated and land use systems. Symposium on Biomass Energy Systems: Building Blocks for Sustainable Agriculture. World Resources Institute. Washington.

Raintree, J.B. and F. Turay. 1980. Linear programming model of an experimental leucaena-rice alley cropping system. *IITA Research Briefs* 1(4):5-7.

Raintree, J.B. and K. Warner. 1985. Agroforestry pathways for the integral development of shifting cultivation. IX World Forestry Congress. Mexico City.

Reddy, A.K.N. 1985. Energy for biomass and biomass for energy: a development-oriented view. Symposium on Biomass Energy Systems: Building Blocks for Sustainable Agriculture. World Resources Institute. Washington.

Rocheleau, D. 1984. Criteria for re-appraisal and re-design: intra-household and between-household aspects of FSRE in three Kenyan agroforestry projects. Annual Farming Systems Research and Extension Symposium. October. Kansas State University. Manhattan, Kansas.

Rocheleau, D. and A. van den Hoek. 1984. The application of ecosystems and landscape analysis in agroforestry diagnosis and design: a case study from Kathama Sublocation, Machakos District, Kenya. Working Paper No. 11. ICRAF. Nairobi.

Ruthenberg, H. 1971. *Farming Systems in the Tropics.* Clarenden Press. Oxford.

Salisbury, R.F. 1962. *From Stone to Steel.* Cambridge University Press. London.

Schirmer, A. 1983. *The Role of Agroforestry in the Pacific.* University of the South Pacific. Suva, Fiji.

Seth, S.K. 1981. India and Sri Lanka agroforestry. FAO/SIDA Forestry for Local Community Development Program. FAO. Rome.

Smith, J.R. 1950. *Tree Crops: A Permanent Agriculture.* Devin-Adair. New York.

Spencer, J.E. 1966. *Shifting Cultivation in Southeastern Asia.* University of California Press. Berkeley, Los Angeles and London.

Stoler, A. 1978. Garden use and household economy in rural Java. *Bulletin of Indonesian Economic Studies* 14(2):85-101.

Torquebiau, E. 1984. Man-made dipterocarp forest in Sumatra. *Agroforestry Systems* 2(2):103-128.

Torres, F. 1983. Agroforestry: Concepts and practices. In D.A. Hoekstra and F.M. Kuguru (eds). *Agroforestry Systems for Small-Scale Farmers*. ICRAF. Nairobi.

Torres, F. and J.B. Raintree. 1983. Agroforestry systems for smallholder upland farmers in a land reform area of the Philippines: the Tabango case study. Working Paper No. 18. ICRAF. Nairobi.

Tosi, J.A., Jr. and R.F. Voertman. 1964. Some environmental factors in the economic development of the tropics. *Economic Geography* 40(3):189-205.

Unesco, UNEP and FAO. 1978. *Tropical Forest Ecosystems*. Unesco. Paris.

Unesco, UNEP and FAO. 1979. *Tropical Grazing Land Ecosystems*. Unesco. Paris.

Verimumbe, I., H.C. Knipscheer and E.E. Enabor. 1984. The economic potential of leguminous tree crops in zero-tillage cropping in Nigeria: a linear programming model. *Agroforestry Systems* 2(2):129-138.

Watters, R.F. 1960. The nature of shifting cultivation. *Pacific Viewpoint* 1:59-99.

Weber, F. and M. Hoskins. 1983. *Agroforestry in the Sahel*. Department of Sociology, Virginia Polytechnic Institute and State University. Blacksburg, Virginia.

Weinstock, J.A. 1983. Rattan: a complement to swidden agriculture. *Economic Botany* 37(1):58-68.

Wiersum, K.F. 1982. Tree gardening and taungya on Java: examples of agroforestry techniques in the humid tropics. *Agroforestry Systems* 1(1):53-70.

Williams, R. 1985. Potential roles for bioenergy in an energy efficent world. Symposium on Biomass Energy Systems: Building Blocks for Sustainable Agriculture. World Resources Institute. Washington.

Wilson, G.F. and B.T. Kang. 1980. Development of staple and productive biological cropping systems for the humid tropics. In B. Stonehouse (ed). *Biological Husbandry: A Scientific Approach to Organic Farming*. Buttersworth. London.

Women, Trees and Tenure: Implications for Agroforestry Research and Development

Dianne E. Rocheleau

International Council for Research in Agroforestry

Introduction

Women's actual and potential contributions to rural development have been widely discussed in recent years. The reciprocal contributions of development to rural women have also received some, though noticeably less, attention (Tinker 1981, 1976a; Hoskins 1979, 1983; Mazumdar 1984; Muchena 1984). The longstanding recognition of rural women's responsibility for basic family welfare (food, water, health and hygiene) has led to a multiplicity of welfare programs focused on women, whether as heads-of-household, wives, mothers or as individuals. Rural women's roles in Third World food production and fuel consumption have also become highly "visible" in recent years (Muntemba 1982, Deere and de Leal 1981, Ki-Zerbo 1981, Brokensha and Castro 1984, Brokensha et al 1983, Hoskins 1983, Williams 1983). The importance of women's labor is now routinely mentioned in agricultural, agroforestry and social forestry circles, but often without a realistic assessment of women's needs, motivations and limitations as clients in their own right (Agarwal and Anand 1982, Williams 1982a,b,). The terms of availability of their labor and its combination with the other factors of production also require careful study and imaginative responses (Tinker 1976b, Nagvi and Gupta 1980).

Some sectors of the rural development community have recognized the need to improve women's access to, and control of productive resources (land, water, capital, information) to enable them to increase and/or stabilize production on their own behalf (Berry et al. 1983, FAO 1983a,b). Where women already have access to such resources, development efforts need to help safeguard their customary rights rather than erode them through external intervention (Colfer 1984). In either case actual progress has been slow and irregular, primarily due to 1) a lack of political will and/ or 2) lack of information and understanding about women's past, present and potential access to, and control over the means and fruits of production in rural areas. The following directly addresses the information gap, in both technical and policy sectors, in the hope of motivating techno-political will by talking in terms of the real and the possible in regard to women, tenure, and agroforestry.

Rural women and agroforestry programs have much to gain from a well informed and well defined association. Tenure is a key point of information, definition and/or action in cases where improved production is the main shared objective. In cases where equity is a major objective, both tenure and technology changes may be used as mutually reinforcing means to that end (Riddell this volume 1985). However, tenure reform and technological innovation will not suffice. Whereas tenure refers to ownership of, control over, or access to space and fixed objects (plants) in space, production also requires time, energy, capital, information, and control over the use thereof.

Numerous experiences in rural development (Papanek 1981, Siedman 1981, Peacock 1984) social forestry (Skutsch 1983), land tenure reform, and special women's projects (Unasylva 1984b) have demonstrated that the road to failure is paved with good intentions, bad information, and inadequate treatment of women's access to one or more of the factors of production. However, a few project histories (Hoskins 1979, Scott 1980, Wiff 1984, Fortmann and Rocheleau 1984) and a wealth of experience in traditional and evolving agroforestry systems suggests that a symbiosis is possible. A firm basis for cooperation rests on careful attention to a few key points:

1) the factors that differentiate women in regard to land tenure, tree tenure, and agroforestry technology;

2) the social units of organization and production which integrate their rights, responsibilities and participation;

3) the landscape and ecosystem units which integrate their physical resource base with the social reality;

4) the desirability and practicality of working within, working around, modifying or radically changing any or all of the above.

Points of differentiation and integration

Whether we consider women apart, or women as a distinct client subgroup within the larger population, the terms of their participation will usually be distinct from those of men, owing to *de jure* or *de facto* differences in rights and responsibilities. This is especially true with regard to the quantity, quality and terms of access to land (Black and Cottrell 1981). Women's access to other productive resources (water, draft power, agrochemicals, labor, information) usually differs from men in degree and in kind. Moreover, control over the components (animals, crops, trees, shrubs, pasture) and the products (food, fodder, fuel, timber, cash, fibre, medicine) of agroforestry systems is often subject to rules distinct from those governing men's actions. All of these differences may be expressed in the division of men's and women's separate places and activities, in nested complementary roles in the same places and activities, or in sharing of interchangeable roles (Boormans 1984, Kelly 1981, Vayda et al. 1980).

While all of the differences cited above may limit the scope and nature of agroforestry technology and project design, there are also distinct advantages and opportunities for agroforestry within women's separate domains of space, time, activities, interests and skills. They may also have special knowledge and special rights and obligations *vis-a-vis* certain categories of artifacts (tools, infrastructure) and natural objects and phenomena (water, fire, animals, plants, animal and plant products).

Likewise agroforestry may imply new demands on women clients: negotiating new arrangements for use and management of shared lands (household or community), new labor or capital inputs, learning new skills, and closer management of soil, water, plants, and animals in existing woodland, cropland, pasture or boundary lands. On the other hand, agroforestry may serve to validate women's land and tree use rights or claims to ownership, to increase production, to decrease gathering time and to reconcile conflicting objectives for shared household or community plots (e.g. from cash crops *vs.* fuelwood to cash crops *plus* fuelwood).

Both the opportunities and constraints of differential roles warrant a closer look at each of those domains where differences count, and at the potential for their integration into a larger, unified system of production and distribution.

Legal status

Women's legal status usually differs from men's in terms of clarity and strength of defined rights and responsibilities, as well as in the nature, scope and complexity of rights and obligations. Even in countries where traditional law (e.g. Moslem law, Mayer 1984) or modern legal reforms have provided for equality in most spheres, strong discrimination often exists with respect to use and ownership of agricultural land (Ahooja-Patel 1982, Mazumdar 1982, Fortmann 1981). This has obvious implications for women's adoption of agroforestry technologies that depend on secure access to land on a permanent or long term basis.

Statutory, civil and customary law

In some cases *de jure* differences in men's and women's land and tree rights are clearly stated in statutory or civil law (FAO 1979). More often these codes defer to (coded) customary law in matters of "marriage" and inheritance, which include most of the cases relating to women's land rights. While civil and statutory law usually applies to all people in all regions of a given state, customary law may vary by region, religious affiliation, or ethnic origin, adding to the complexity of interpretation and administration. Within both civil and statutory law, as well as customary law, women's rights are often tied to marital status (FAO 1979) though the most independent group under one customary code may be the most powerless under another body of customary law (married *vs.* widowed, divorced, *vs.* single). The result is a complex array of types of rights within parallel legal codes, nested within units ranging from the individual to the kin group.

In addition to the multiplicity of systems, the process of codification and interpretation of customary law is itself riddled with mistaken and/or arbitrary judgments in reconciling culturally incompatible categories of rights, obligations, property, and units of social organization. Whereas many outsiders are quick to point out the discriminatory nature of traditional systems, it is often in the translation to "modern" legal terms that women lose out (Onger-Hosgor 1983, Muntemba 1982). In Tanzania and Malawi the customary law was standardized at the national level after independence, resulting in a uniform patrilineal system. This deprived the women of matrilineal groups of former land use and residential rights in their own villages by allocating land to households through male heads (Fortmann 1981, Brain 1976).

Land tenure reform

In other cases it is the interpretation of the customary law during the implementation of land reform and settlement programs that undermines women's access to land. One well documented example illustrates this process:

> Luo women in Kenya lost their rights and security of access to lineage land (in the commons and in the household plots) when land adjudication officers equated the inherited right to allocate land with land ownership, ignoring all other rights. Male heads of household and male relatives of women heads of households were alloted exclusive rights of ownership, with few exceptions. By contrast, the actual categories of traditional rights included ownership (vested in the lineage), the right to allocate land (vested in married men), membership in the parilineage (passed on to sons through the mother), and rights of usufruct and residence, including access to common lineage lands as well as to own plots within husband's allocation (acquired by women through marriage) (Okeyo-Pala 1977, 1980; Okoth-Ogendo 1981).
>
> The Luo women did not lose their access to all the lands they previously used, but they have lost some lands and all *security* of access. They have also lost the status they formerly held as channels of lineage land to sons. With each succeeding generation women may be less able to rely on tradition to protect their intra-household autonomy as farmers and their access to *de facto* communal lands as gatherers (Okeyo-Pala 1980).

Not only does this case exemplify the loss of prior customary rights, it also points out the elements of ambiguity, dependence, and insecurity that replaced a fairly definitive, complementary, and stable relationship between men and women *vis-a-vis* land. While the men's rights and inter-household distinctions were well defined relative to the legal system, women's rights and intra-household distinctions were left in the realm of customary law, as modified and circumscribed by the new tenure system. This opened a wide gap between *de jure* and *de facto* rights and responsibilities and made women's farming and gathering activities dependent on permission to use individual men's property.

This example is by no means isolated, nor does it exclusively represent any one culture, region, or form of political economy. The erosion of women's often substantial customary rights under modern legal reforms (colonial and post-independence, capitalist and socialist) is a widespread phenomenon (Onger-Hosgor 1983, FAO 1979, Cultural Survival Quarterly 1984a). It has been reported as an accomplished fact for many ethnic groups throughout Africa (Fortmann 1981,

Barnes 1984, Afonja 1981, Muntemba 1982), Asia (Nowak 1984, Endicott 1984, Cultural Survival Quarterly 1984b, Palmer 1978, Mazumdar 1982) and Latin America (Arizpe 1982, Deere and de Leal 1981, Mickelwait et al. 1976). Sometimes this has been done on the pretext of formalizing customary tenure systems, as above, and in other cases it has simply been imposed by colonial powers or by national governments on ethnic minorities. This was particularly pronounced during the colonial period with respect to matrilineal groups, extending to de-registration of women's ownership or use rights in Zimbabwe, Zambia (Muntemba 1982) and Malaysia (Boserup 1970).

The tendency of most land tenure reforms in agricultural and mixed farming areas has been to expand and legalize men's access to land, while decreasing and "informalizing" women's rights of ownership and use. Rights of usufruct in general, and overlapping multiple use rights in particular, have been largely ignored in favor of individual or state ownership. In both cases the titles or the allocations of land are made on a household basis to male heads, or to associations whose members are exclusively or predominantly male heads of households. Even where a separate commons is recognized as part of the reform process, as in India and Pakistan, unregulated overuse and privatization by influential individuals (Cernea 1985) reflect the inadequacy of legal and administrative infrastructures in maintaining women's access to land and trees.

These trends do not bode well for women's land and tree use rights, but the process is just beginning in some areas and has not yet reached others. There are also some examples of women (Yoruba and Ghanian) gaining *more secure* rights to land after privatization (Afonja 1981), although the questions then arise "how many?" and "which women?"(Mazumdar 1984). Some governments have implemented reforms-of-reforms, as in Mexico where government changed the rules of membership in Ejidos to allow women to retain their own plots after marriage (FAO 1979). Some recent land reform programs have also incorporated special provisions for sharing of benefits by women while others have simply perpetuated prior *de facto* exclusion of women from land ownership and management (Onger-Hosgor 1983). In other cases landless women have gained access to land through revolutionary or liberal land reforms, although the terms of access are limited by marital status, are administered by male relatives, or are vested in associations of male heads of household, as in Ethiopia (Onger-Hosgor 1983, Whalen 1984).

Gap between women's *de jure* and *de facto* rights

In addition to formal legal differences between men's and women's land and tree rights, there is often a widening gap between women's *de jure* and *de facto* rights and responsibilities. Land reform programs have generally formalized and strengthened men's individual ownership, at the expense of women's control of lands vested in the lineage or at the expense of their well defined rights of usufruct. The latter often persist as informal or *de facto* rights, but with no meaning or value in the formal sector where contracts are signed, land is bought and sold, credit is disbursed, and technical assistance is given. As beneficiaries of land reform, men's legal rights often include and exceed former customary rights, and as such men have less need or interest as a group in recourse to customary law. Women's rights, by contrast, are either nested within rights legally ascribed to male relatives and community groups, or they are part of a growing body of uncoded, evolving customary law, outside the formal legal system.

The disparity between *de jure* and *de facto* rights in many areas is a land reform crisis in the making, particularly for women, but also for landless people in general. It may well be that rights of residence, and informal access to land, however insecure legally, are what keep many women and landless people in the countryside. For many women the inability to dispose of property is of little consequence, as long as they have farmland to work. However, as urban-dwelling sons and other relatives inherit and take possession of the land, they may change the terms of access (Haugerud 1983) as well as the land use and land cover.

While rights of residence and gathering may persist, the land use changes may drastically reduce availability and quality of tree, shrub and crop by-products. The legal heirs may sell multipurpose trees for timber or charcoal, unless women residents have legal tree ownership or well-defined rights of usufruct. For example, in some groups on the Kenyan coast, widows retain exclusive rights to harvest fruits and nuts from trees, while the land itself may be inherited and managed by another relative. Women's tree rights are also recognized among the Ibo of Nigeria and among Tanzanian coastal people observing Islamic law (Obi 1963, Tanner 1960, Fortmann this volume).

Where more egalitarian or separate-but-equal rules now apply (as in some groups of pastoralists, forest dwellers and hill people) women may soon lose those rights under poorly conceived land tenure reform or land settlement projects. In many areas development projects have set precedents for subsequent survey and land allocation/registration procedures by dealing only with male heads of household and community leaders (Nowak 1984, Endicott 1984). In other cases, extension services and development projects have mistakenly blamed matrilocal land tenure systems for low adoption rates of agricultural innovations (Kishindo 1984), calling for introduction of patrilocal and patrilineal systems.

Strengthening or improving women's tree ownership and rights of usufruct might well provide a means to reconcile women's needs for secure access to trees and tree products with men's legal ownership of the land. Where such rights are not well-defined there may be a gradual shift to men's *vs.* women's land use types, and men's *vs.* women's trees in the landscape, as legal distribution of land and trees is increasingly reflected in actual use and management.

The causes of these trends and of the widespread imbalance between rural women's needs, responsibilities and rights have been, and continue to be, hotly debated (Arizpe 1982, Deere and de Leal 1981, Afonja 1981, ICES 1985, D. Jain 1984, Mazumdar 1984). Varying degrees of blame have been placed on traditional, colonial, neo-colonial, nationalist, capitalist and/or socialist political economies. *The fact is that rural women's responsibilities are often way out of balance with their legal status and formal rights to land, trees, and the products of both.* Discrepancies between women's rights, responsibilities, and authority, between *de jure* and *de facto* rights, and between coded and uncoded customary law, all must be faced by agroforestry and social forestry planners and project personnel, as well as by land settlement, land tenure reform, and women's program specialists.

There is ample scope for agroforestry and social forestry projects to set positive precedents in the recognition and creative translation of women's land and tree rights into coded contractual forms at the project or local administrative level (Hoskins 1979, 1983).

Division of space, time and responsibility

In most rural areas of the Third World, men and women have distinct domains of activity, responsibility, control and knowledge. The extent of this division varies from almost nil, as among the Dayak forest dwellers of East Kalimantan (Colfer 1981) to very strictly defined separation of men's and women's domains, as maintained in some parts of India and the Middle East, particularly where purdah and other forms of seclusion are observed (Ahooja-Patel 1982). The degree, type and terms of division vary substantially between regions, ethnic groups, and classes, based on differences in customary law and practice.

During the past few decades most groups have also experienced dramatic changes in the division of space, time, expertise, and authority, due to cash cropping and other commercial enterprises. The effects vary in quality and degree, as illustrated in the following examples:

1) Virtual domestic unemployment of Rashiidy Bedouin women, whose former home construction and fuel gathering tasks are obsolete in the cash economy of new settlements (Young 1984).

2) Displacement of Yoruba women's food crops by men's cash crops on prime cropland, followed by women's increasing responsibility for food crops and entry of some women into cash cropping enterprises as land and/or tree owners (Afonja 1981).

3) Women's assumption of men's labor and management roles where men have turned to off-farm labor in cities or in cash crop estates, as in Kenya (Okeyo-Pala 1980) and Peru (Arizpe 1982, Deere and de Leal 1982), or in distant mines as in Lesotho (Gordon 1981), Botswana (Fortmann 1984), Zambia (Muntemba 1982) and Bolivia (Harman 1984). These cases of men's migration do not imply a homogenization of roles, but rather a new division of space and labor on a different scale or on different terms.

Whether clients are following long-established customary practice or recently adopted role definitions, men and women in any given household or community are likely to have distinct interests, needs and capabilities *vis-a-vis* agroforestry or social forestry. Division of labor, expertise, authority, and spatial domains will influence the participation of women in new types of technology and land use systems. For settlement or agroforestry projects committed to women as clients, these role differences can also determine: the range of landscape niches available; the potential spatial arrangements of trees, crops, pasture and animals; the types of trees and tree products to be included; the types and amounts of labor and management inputs; and the types of contract or special tenure arrangements needed.

Spatial niches in the rural landscape

Visible landscape patterns and features are an excellent point of departure for determining the spatial distribution of men's and women's domains and potential niches for shared, separate or interlocking agroforestry technologies. Given the cultural and environmental diversity of land use systems and the dynamic nature of community development cycles and land use change, little can be assumed as to which niches will be used, managed, shared or owned by women. While live fences are the major opportunity for women's agroforestry technologies in some parts of western Kenya, the external boundary fences are the exclusive domain of men heads of household among neighboring groups. In some areas women still manage separate food and cash crop plots of their own, and in still others men and women tend and harvest separate plants within the same multistoried agroforestry systems.

While there are no niches universally used and managed by women, there are some spaces which are more often their domain. Strangely enough, the two niches of greatest importance to women are often the closest to home and the farthest away, respectively. Home gardens are located near the center of household activity, and common gathering areas (forests, bushland and grassland) are usually peripheral to the home and croplands or to an entire settlement, depending on population density and land use intensity (Raintree and Warner 1985). While the first (intensive land use) is located so as to minimize the opportunity cost of time away from the home, the location of the second (extensive use) minimizes opportunity cost of land and actual labor and management inputs on-site. A closer look at both of these land use types, their relative position in the landscape, and their importance to women, provides insights into general considerations *vis-a-vis* spatial and functional niches for women's agroforestry technologies and related needs for tenure and technology innovations.

Home Gardens. The cultivation and management of home gardens by women is a widespread phenomenon among settled groups the world over (Buch 1980, Ninez 1985). This is particularly pronounced in Latin America in areas where women do not traditionally till the land, since it provides an agricultural production niche that is seen as an extension of the home. The home garden is often a way around taboos against tilling the main cropland, and is usually considered an extension of the home as the women's domain. Moreover, by definition such plots are location-specific to the home area, and as such are accessible to women whose mobility may be limited by custom, or by the complex logistics of mixing travel with child care, food processing and food preparation. Home gardens provide an opportunity to intensify labor inputs to increase production, without adding time away from home and within a flexible schedule shaped around other household responsibilities (Chaney and Lewis 1980).

While home gardens may occur within systems ranging from shifting cultivation to intensive multiple-cropping in permanent plots, they seem to be the domain of women wherever such a plot is one among many other plots available to the household, or in cases where men are almost exclusively engaged in off-farm labor. In intensively cultivated areas of acute land scarcity, the whole household may work the home garden under the management of the head of the household, as in parts of Southeast Asia (Sommers 1978, Hunink and Stoffers 1984). In such cases, the rationale for the home garden shifts more toward labor intensification on scarce land rather than efficient multiple use of women's scarce time. Even so, these plots may have greater relative importance for women than for men, based on the distribution of labor input and on the fact that men may have alternative sources of cash income (Stoler 1978). This is also reflected in the tendency for women in Java to inherit home gardens, while their brothers inherit the rice croplands (Palmer 1978, FAO 1979).

The home garden is uniquely suited for agroforestry projects with women. The limited plot size encourages multistoried systems, while the woman's *de facto* control and the permanence (or relative permanence) of the site encourage investment in tree crops and site improvement (terraces, manuring, fencing). The small plot size also implies a high ratio of peripheral to enclosed area (Rocheleau and Hoek 1984) and hence a relatively high proportion of the site production potential could be allocated to multipurpose living fencerows. The site can also be an ideal place for small livestock, such as chickens or caged rabbits, and may provide residues useful for feeding hogs or goats confined nearby, or supplementary fodder for a larger milk animal. Likewise, such an intensive small plot is often a cost-effective site for application of manure from livestock confined nearby.

Although the home garden is primarily a phenomenon of settled agriculturalists, there is an analogous niche among some semi-nomadic pastoralists. Milk animals, young animals or sick animals may be kept by women in special small corrals close to the home, as among the Maasai in southern Kenya (Barbara Grandin personal communication, Penny Nestel personal communication, Nambombe 1984). As in the case of the home garden, the location of these special corrals minimizes time away from home and allows for a more than usually intensive investment of time, attention and protection. Such a land use unit constitutes a reasonable basis for introduction of a fodder-based agroforestry technology around and adjacent to it. By providing high quality fodder on-site, such a technology would reinforce the ideas of intensive care, convenience to the home, and milk production for sale or distribution by women.

Some features of home gardens may also be applicable to community or social forestry approaches (Chakravorty 1980), particularly the features of location near the home and of species and product diversity. If fuelwood or tree products are the main objectives and women have landscape niches available close to home, they may be better able to plant and utilize such trees in a home garden or in a block planting of equivalent size close to home. Time, distance and distribution of costs and benefits may discourage some women from participating in large community plots. This does not automatically imply a household level organizational approach (e.g. for seedling production in nurseries), but simply an on-farm planting niche similar to the home garden.

The other aspect of home gardens that might be partly integrated into social and community forestry is their structural and functional diversity. Women's groups in some projects have specifically requested the project to combine vegetable gardens with tree nurseries (Rocheleau 1985), particularly where sufficient water was not available at the household level, but was available at a community site.

This idea could be extrapolated to interplanting of trees and vegetables. Multipurpose trees combined with horticultural trees and vegetable crops in a large community plot may be more useful to women than a single-species stand of timber or fuelwood trees. Moreover, timber or fuelwood species compatible with such a multistoried, multipurpose system could help to alleviate conflicts between men's and women's trees, as in Kakamega District, Kenya (Chavangi 1984), or men's *vs.* women's land use, as in Kisii District, Kenya (Fortmann and Rocheleau 1985, Barnes 1984) and between commercial trees and subsistence agricultural land use, as reported for some social forestry projects in Gujarat, India (Foley and Barnard 1983).

Establishment of valuable horticultural crops in the plot can also help to identify the site as one of value, thus strengthening "social fences" against grazing and browsing. While trees are respected as property in some places, herbaceous crops are more valued and respected in others. Although the problem might shift from one of damage due to neglect to losses due to theft, the participants might well be more motivated and capable of solving the latter.

The Commons. While the commons is rarely the exclusive domain of women, it is often a major source of subsistence and commercial products for women. In some stages of land use intensification and landscape development it may become the major source of women's livelihood or of their contributions to the household and community (Roy 1980, Dahlberg 1981). Where land is plentiful, as among some forest dwellers, shifting cultivators and pastoralists, the forest or rangeland is a shared domain of men and women, with division more likely to occur on the basis of labor, expertise and security. Women are usually, but not exclusively, responsible for gathering food and fuelwood, as well as fibre, some medicinal plants and other "minor" forest and range products. They may also manage grazing or browsing animals, but are generally more responsible for the other products and activities listed above.

The communal grazing and gathering areas may be differentiated from the household lands by use alone, if at all. However, this domain deserves special attention *vis-a-vis* establishment of tree ownership and land use rights early in the process of land use change. While men may replace their foraging activities with wage labor or intensified agricultural and livestock production, the group may continue to rely heavily on forest and range products gathered by women. Safeguarding or expanding women's tree ownership and rights of usufruct in surrounding forests and rangeland may help to prevent environmental degradation and to maintain important sources of food and other products, as well as maintaining women's status and customary rights to use and protect forest and range lands of adequate extent and quality.

As village development cycles and land use conversion and intensification proceed, the commons may become a residual domain, left to women by default. In the process of land use intensification from bush fallow to multiple cropping (Raintree, this volume) women's access to and dependence on gathering areas may change substantially, and the commons itself may shift from one spatial niche to another over time. This transition is also marked by constant adjustments between use of the commons and use of interstitial niches on-farm.

In bush fallow systems women gather from forests at the village periphery, often in combination with work trips to outlying agricultural fields of the lineage or household (Ruthenberg 1976). Many of the same products or their equivalents are also gathered from scattered fallows in various stages of development. As the length of fallow is shortened in response to land shortage/population pressure (Boserup 1970, Lagemann 1977), the diversity, quality and quantity of these products changes substantially. It is at this point that women may intensify their gathering activities in the peripheral forest or in croplands and short fallows, depending on the distance to the forest, the type of products sought, and the terms of access to forest and farm lands.

Women may eventually find their access to household or lineage fallow lands preempted by permanent cropping, and even their access to seasonal fallows may be curtailed by multiple cropping and/or irrigation. Changes in crops and reduction of crop and weed diversity can remove valuable sources of cropping system by-products. This process has been documented among landless gatherers in some parts of India (Das Gupta 1984). It has also been reported, in various stages of development, by women in Kenya (Barnes 1984, Fortmann and Rocheleau 1985) the Dominican Republic (Rocheleau 1984a) and Zambia (Huxley et al. 1985). Unless gathering remains an activity of the landowners themselves, its products are liable to be relegated to the status of by-products with no niche of their own, unless the plants or the niche are re-defined as gatherers' property. Perhaps this is what accounts for the integration of forest form and function into traditional multistoried agroforestry systems in areas where gathering was shared by men and women, and the intensification process was guided by the traditional values of both.

Agroforestry technologies can either reinforce or ameliorate this problem (Hoskins 1983). For example, a new alley-cropping system could displace shrubs or weeds that provide leafy vegetables (Fortmann, this volume). On the other hand, enrichment planting and more intensive management of selected (exotic or indigenous) woody species in cropland and grazing lands could provide forest or range products for women within a system compatible with men's crops or pasture. This would, however, require a clear statement of women's tree ownership or rights of usufruct.

As farmland use tends increasingly toward monoculture or more uniform and intensive cropping, women may rely increasingly on distant forest or bushland as sources of gathered products. They may also opt to settle for the basic essentials (fuel and fodder) that can be gleaned from a degraded, overgrazed commons, sacrificing quality, quantity and diversity of products for proximity. Some women modify their schedules to fit the spatial distribution of resources, with frequent visits to nearby commons and occasional visits to more distant forests for special products or for goods seasonally scarce in the farmland and commons (such as tree fodder). The latter is widely practiced in the Himalayan foothills of India (Bhatt 1980, Mishra and Trepathi 1978, Raintree et al. 1985), though gathering pressure on the forests is increasing as the commons are degraded or converted to cropland. A similar process of slow drift from government forest to Panchayat commons and from commons to private land has been observed in parts of Pakistan (Cernea 1985).

In areas where a commons does not exist as such, women often rely on the "borrowing" of stickwood, grazing privileges, and other non-commodity goods and services from private woodland or pasture lands of a neighbor, relative or patron. This practice is widespread in Kenya among neighbors and clan members. In Latin America the patron-client relationship prevalent in rural areas often includes such arrangements.

As in the case of land use changes at the farm level, women-as-gatherers and/or herders may find their communal and "borrowed" sources of forest and range products reduced in quality, extent, and accessibility. Just as the gathering grounds on-farm may have been "up-graded" to more intensive uses, the village commons can be annexed by influential members of the community (Cernea 1985, Foley and Barnard 1983) for cropland, grazing land or "anti-social forestry." However, the commons and nearby forest are often overutilized to the point of severe degradation or are reduced to residual remnants of successional vegetation in fragile environments. In some cases the frontier of the commons and the forests have been pushed to the limit (Roy 1980), as defined by women's walking distance, or by their tolerance for conflict with neighboring villages or government authorities.

Women's tree and land ownership and rights of usufruct for common lands may apply to individuals, households or groups. However, the issues of group ownership and distribution of rights and products between group members are likely to be more important than in on-farm approaches. While the "tragedy of the commons" is an oft-invoked (and perhaps overused) concept, there are several points which favor helping women consolidate and maintain well-defined, conditions and terms of tree ownership or rights of usufruct in village commons, in public forests and rangeland, or in "unclaimed" interstitial niches such as roadsides, gullies and boundaries.

The advantages of commons areas for women's practice of social forestry and agroforestry include :

1) Locational advantages such as proximity to markets, water sources or other central places, as opposed to distant or scattered gathering areas ("one-stop shopping").

2) Unique site quality such as bottomlands, riverbanks, ravines, or special soil types or vegetation with site-specific resources not widely available on-farm.

3) Economies of time with respect to the amount and flexibility of time spent at site (if cooperative labor rotation is arranged for convenience of members and not vice versa). The relative convenience of numerous small plots near the home *vs.* consolidated group plots will also depend heavily on local experience with labor exchange and labor sharing arrangements.

4) Minimum size for establishment or maintenance of specific ecosystem types may be more suited to large community plots.

5) Economies of scale *vis-a-vis* fencing, site preparation, protection, labor for maintenance and marketing of products.

6) Availability and ease of access relative to farm lands (especially important for landless women or those with no access to separate plots or no freedom of experimentation on household land).

7) Accessibility of training and assistance and advantages of "group learning curve" (Wake 1984) if commons site is managed by a group.

8) Women may gain separate access to credit, individually or as a group, based on shared title to a site (legal feasibility varies).

While home gardens and the commons do not by any means exhaust the potential niches for social forestry or agroforestry with women, they do illustrate the interplay of land use change in the larger system with women's access to land and trees. They also illustrate that women's involvement in decision-making and land use planning (at farm or community level) may be the key factor in choice of niche type, site and technology for women clients of agroforestry and social forestry programs. Tree ownership and/or contractual agreements as to rights of usufruct may help to maintain or establish diverse, managed agroforestry systems, whether integrated within plots, farms, watersheds or other landscape units.

Division of labor, expertise and authority

For women to take full advantage of spatial opportunities in the landscape, consideration must be given to the sexual division of labor, expertise, and authority, including specialization by enterprise, activity, plant species, plant parts, management techniques and products.

Labor, expertise, and interests. The degree and significance of the division of labor between men and women varies as much or more than access to space. In fact, it is often a major determinant of spatial and ecological specialization. Some of the main ways in which men's and women's labor is divided include:

1) Complementary specialization in different subsistence activities, such as: hunting *vs.* farming, herding, gathering; clearing *vs.* planting (in the same plot for the same crops); planting and tending of different plants for home use; production *vs.* processing of goods for home consumption.

2) Division of commercial *vs.* subsistence activities: cash cropping *vs.* subsistence cropping, gathering or herding; commercial *vs.* subsistence fishing, gathering, hunting or herding.

3) Division of activities within own cash cropping systems: land clearing *vs.* planting, weeding, harvesting; production *vs.* processing or marketing; management *vs.* labor.

4) Division of employment categories: wage labor *vs.* self-employment *vs.* unpaid household labor; casual vs. seasonal *vs.* permanent off-farm labor; skilled *vs.* unskilled off-farm labor; nearby *vs.* distant off-farm labor; agricultural *vs.* non-agricultural labor.

All of these forms of specialization have implications for the advisability and practicality of separate, joint, or interlocking men's and women's agroforestry technologies and programs. Many of the divisions listed above imply cooperation, competition, or coexistence between men and women *vis-a-vis* land and trees. For example, in Kakamega District, Kenya there are taboos against the planting of trees by women (Chavangi 1984), which also coincide with prior concepts of planted timber trees as part of men's domain *vis-a-vis* house construction. The division is further reinforced by the commercial value of timber, coupled with men's dominance in cash cropping and commercial enterprises. The prohibition against "tree" planting by women will probably not apply to *Sesbania sesban* and other small fuelwood and alley cropping species. Similar examples abound in regard to distinct men's and women's priority use values stated in diagnostic interviews, extending to rigged voting, arguments and negotiation in group interviews over the relative merits of fodder *vs.* fuelwood trees, or fodder *vs.* timber (Rocheleau 1984, R. Vonk personal communication, Hoskins 1983, Joshi 1982, Fortmann and Rocheleau 1985).

Although the subsistence vs. commercial conflict in men's and women's interests (Sharma 1981) has received the most attention, there are often differences between preferred pathways to commercialization, reflecting the relative importance of men's and women's skills, labor, land resources and opportunities within alternative enterprises or development strategies. The potential for women's participation in commercializing minor forest products *vs.* timber resources in the Himalayan foothills is a case in point (Surin and Bhaduri 1980). Women are already interested and involved in cash enterprises based on gathering, processing and retailing of many forests products, and might be best served by projects to improve and sustain that activity rather than planting new stands of trees that will not yield products for processing (i.e value added) by women. Moreover, enterprises based on "minor" forest products depend largely on renewable resources. This is yet another possibility for agroforestry systems that address women-as-gatherers while intensifying overall production in both cropping and forest-based systems.

The scale at which interests are divided will also depend upon the extent to which off-farm labor and migration have influenced the division of labor and management (Kerven 1984). In many areas men's and women's spatial domains of work may be split on rural/urban lines, or estates *vs.* smallholdings. Labor may be split according to wage *vs.* unpaid labor or self-employment, as opposed to previous divisions based on particular operations or skills within the small-holding. Where their labor is required for well-defined seasonal tasks, as with plowing, men may still retain this role in their households (Fortmann 1984). Authority may still be vested in the male heads of household, with varying degrees of freedom for women managers to make management decisions about land, trees and land use change (Cantor 1984, Caplan 1984, Rocheleau 1985). The number and proportion of such "statistically invisible" *female-managed* households has increased dramatically in many countries, and in Kenya these outnumber female-headed households by a ratio of 2:1 (Hunt 1984).

Female-headed households may also constitute a substantial part of the client group (Buvinic and Yousset 1978), since they account for roughly 24 to 33 percent of all households in the world. While women in such households may theoretically have more freedom of decision and less division of rights and responsibilities at the household level, all but the wealthiest or largest female-headed households may be disadvantaged by lack of special male skills, an absolute shortage of labor and loss of off-farm income (Fortmann 1984, Jiggins n.d.), as is the case among the Bemba of N.E. Zambia. There, most female households shift production emphasis from the *chitamene* rotation of grains, tubers and legumes, to cassava monocropping in permanent plots, because they require male labor for the land-clearing and wood-cutting stages of the *chitamene* cycle (Stolen 1983). Female-headed households and sub-households may also be affected by divisions practiced and enforced at the community level, and they may be disadvantaged in their access to male decision-making and enforcement groups at the community level.

The latter is part of a larger issue of women's variable access to management and decision-making processes within units organization of different scale (household, group, village, or national level). Aside from legal status, which has been treated above, and the day to day division of labor, there is the crucial role of administrative and organizational units in interpreting and enforcing women's access to trees and land, and their preferences with regard to agroforestry technology.

Authority and decision-making. The exact form and degree of female decision-making power and authority varies substantially from one culture (or ethnic group) to another, as well as between sub-groups, between different units of organization, between different domains of authority, and over time within groups (due to social change). In spite of this diversity, there are some common issues to be considered that are relevant to most cases. Moreover, there are many common constraints and opportunities for agroforestry, social forestry and land tenure programs based on the sexual division of decision-making and authority *per se* and the sexual differentiation of how such power is *expressed*.

In many cases male and female authority and decision-making powers are interlocked and nested in a complex system of household, kin group and community governance which recognizes sexual division of skills, knowledge and interests of the type already discussed above. For example among the Luo in Kenya, while men as a group always had more authority over the allocation of land and now have even more control as individuals, through title deeds, women still retain ownership of the harvested crops from their sub-plots and have full authority over the disposition of the grain in their stores and the profits from its sale (Okeyo-Pala 1980, Chaiken 1985).

However, authority and decision-making are often viewed by outsiders as restricted to a legalistic realm of formal rule which they interpret from limited experience. The latter often includes instantaneous "decisions" taken in their presence by men representing households, groups and communities, or other indications of "absolute power" exercised by these same representatives over the rest of the population.

While such conditions may exist in exceptional cases this impression is often a mistaken one, a result of confusing *spokesmen* with *decision-makers*, and immediate *answers* to outsiders' questions with *decisions* that carry the weight of authority for a given social unit. When a decision is really required and taken seriously, many groups will request time to discuss the matter fully, both formally and informally, among all members.

Even apparently concrete indicators of men's control over certain types of transactions, such as cash expenditure, may be misleading. For example in Fakot village (Tehri Garwal District in the outer Himalaya, India) male heads of household make all or most purchases at local markets and seem to have nearly exclusive right to dispose of family income. However, critical questions about women's own interest in increasing household cash income in this same region elicited explanations of how women's "budgets" for household spending do increase with increase in total household income, and that the women, in fact, usually "make the list" (figuratively or literally) for that part of their husband's purchases which corresponds to their own domains of control (e.g. food, certain types of implements, clothes, medicine and other household goods). Similar processes of nested household decision-making were also described with respect to choice of crops and decisions to change the overall cropping system (Raintree et al. 1985). In this case men do seem to have more control in both domains, and are the ones who voice such decisions to outsiders or who discuss them with other men in exclusive male community meetings. However, they are not the sole decision makers and may defer to their mothers and wives on matters of special interest to women. The degree to which decision making is divided or shared, relative to particular domains of interest, skill, and space, may determine in large part what types of agroforestry and social forestry activities should be tried, in what landscape niches, and whether with women clients or with mixed groups of men and women. (Hoskins 1983, Fortmann and Rocheleau 1985, Williams 1985).

Another critical point is the choice of organizational units to mobilize public participation and decision making, or to liaise between the whole client group and research and development programs (Max-Neef 1982, Thomson 1983, Weber and Hoskins 1983). Much of the basis for evolving common law, as well as for land tenure reform and environmental law, rests with a variety of decision-making groups at all scales of organization. Aside from the legal precedents set at an administrative level, there are also countless decisions about land management and land use planning made by groups ranging from the Forest Department to peasants' associations to councils of village elders and kin groups.

The scale and type of social organization chosen for program-client liaison can radically affect the results (Chambers 1983). Thus, the results of land tenure reform among the Luo in Kenya, reflected the absence of women among the adjudication officers, as well as the local advisory committees (Okeyo-Pala 1980). Likewise the decisions of village Panchayats in the Indian Himalayas quite often show a strong bias toward male interests in the disposition of the village commons and in the terms of village participation in development projects, although women have begun to demand membership and a voice in the Panchayat (S. Jain 1984, Berreman 1984, Bhatt 1980, Mishra and Trepathi 1978). In many forestry and agroforestry projects species choices, planting niches and management plans are made by outside technical advisors in consultation with male heads of household or local headmen in the community, resulting in loss of information about women's ongoing work and their knowledge of particular environments and plants and their use and management (Hoskins 1979). Women's involvement in a Honduran agroforestry project through local women's groups has, however, made a positive difference in the success of a new technology (Wiff 1984), as well as in its impact on the eventual disposition of the land and trees currently owned, used or managed by women.

To have any widespread impact on women's participation and benefits it is important to secure their integral involvement in the planning and management of agroforestry and social forestry strategies, programs, projects, and technology development, through organizations willing and able to represent their interests and to facilitate their role in decision making from the national to the household level (Lowdermilk and Laitos 1981). Women clients may have a greater or lesser voice in land and tree-related decisions in household, group or village units of organization. This will depend on the culture, the particular place, and the degree of urban and national influence on local practice.

While many researchers and practitioners of social forestry are disillusioned with the undifferentiated community as a unit of organization and with the commons as a unit of design and management (Cernea 1985), the undifferentiated household is no panacea, either as a unit or organization or as a unit of design and management. While Palmer (1978) states that the "gravitational center of development is shifting to the cooperative, the farmers' association, and the village unit organization," in other areas these structures are breaking down and households are becoming the strongest unit of organization. In yet other cases clan groups or small women's groups are the predominant units of organization for rural development (March and Taqqu 1982, Wijngaarden 1983a, Rocheleau 1985). Moreover, it is not only the strongest unit of organization that is of interest, but also the one that gives most voice and scope to women. The appropriate unit of organization will vary from place to place and with the type of technology and tenure question involved, and whether we are considering women as priority clients or women as part of a larger client group.

The relative advantages of various units may be structurally determined and the choice cf unit may then be critical with respect to women's future participation in land and tree tenure and management programs. However, a particular organizational unit dominated by men may be very stable and well-defined, yet flexible enough to accommodate changes in procedure or composition that would facilitate women's participation.

Although the choice of units will vary widely between cases, the issues involved in that choice will be much the same: whether to strengthen and broaden existing male-dominated institutions and groupings by including women and their interests in collaborative programs (agroforestry, social forestry, tenure reform), or whether to strengthen and diversify institutions or informal groups that already provide ample scope for women's participation/control in decision-making. The criteria for choosing between these approaches will vary with the relative emphasis given to equity and production in any given program, as well as with the predisposition of program personnel and clients to structural versus functional transformation of institutions. In either case there remains a second series of choices as to how to accomplish a reorientation of men's resource management institutions to women's participation, or of women's institutions to new land and tree management issues, activities, and enforcement roles. In yet other cases (as in much of Southeast Asia) the question may be how to protect and strengthen existing egalitarian institutions (Michael Dove personal communication) responsible for allocation and/or management of natural resources. As in the case of other topics previously discussed, this implies a need for new types of policy research and development activities in agroforestry, social forestry and land tenure reform programs.

Indications for policy, research and development action

The problems and opportunities inherent in the sexual division of labor and access to land, plants and their products require special consideration and types of action not yet part of the mainstream approach to agroforestry and social forestry programs. The implications of these differences extend to the content of *technology designs* and *social contracts* for management, as well as to the way research and development activities are carried out with women clients. The gender-based differences in legal status, use of and access to space, types of activities, and control of labor, all have a direct bearing on *what* type of plants can be *planted*, managed, used and harvested, *where, by whom*, for *what purpose* and for *whose benefit*.

Policy formulation

If women are to gain or maintain secure access to the factors of production and the products of agroforestry and social forestry systems, policy interventions may be required on five key points:

1) The very concept of tenure security, as usually applied, requires a broader approach if women's concerns are to be addressed. Nested "bundles" of rights (Riddell, this volume) and interlocking and parallel multiple use of the same spaces and plants must be accommodated within the legislation and procedural regulations that define people's access to land and plants. Since women's rights so often are nested within a larger family or village structure represented by men, the failure to treat the complexity of multiple rights and multiple uses will usually leave women with no formal legal rights to land or trees. By contrast, there is ample scope in some cases to use legislative and regulatory powers to recognize and formalize the separation of land and tree tenure, and to refine and formalize *de facto* use rights and access to trees. Such laws and regulations could provide scope for detailed, locally defined codes to separate rights into complementary domains by assigning specific times, places, types of plants, types of products, or types of activities to particular users and groups of users.

2) Gathering as an activity, and gathered products also warrant recognition within the law and the regulations that define valid uses and users of land. The role of such activities and products in subsistence and commercial economies may need to be documented, in order to justify such formal recognition, but in many cases the information is at hand (Livingstone and Zamora 1983, Lee 1979, Ayensu 1983, Williams 1984, 1985, Hoskins 1979, Surin and Bhaduri 1980, Cernea 1982, Christophersen 1981, Weber and Hoskins 1983, Flores-Paitan 1985) and has only to be interpreted and acted upon. This is a concern of landless people in general, as well as of most women in farming, herding, and forest-dwelling communities (Unasylva 1984b, Cultural Survival Quarterly 1982). Where rights of access can be orchestrated to accommodate gathering and to reserve especially fragile or valuable trees and source areas for exclusive use by gatherers, environmental, equity and production objectives can be combined to the mutual advantage of women gatherers and the community at large.

3) National and regional policy initiatives can also increase women's access to the products and benefits of agroforestry and social forestry systems by promoting women's involvement in, and control of, processing and marketing. The same result could also be achieved by technology designs and market development programs that focus on commodities already controlled by women, or on products with a high potential for value added by women's labor. Such an approach could transform a currently marginal activity with no security, into a major, secure source of income for women. However, further legal or regulatory provisions might be required to maintain women's control, and to protect them from expropriation of newly profitable enterprises into men's domain or into the hands of local elites (men and/or women).

4) Women's ability to secure rights of land and tree ownership and access is also conditioned by their overall legal status, and by the conflicting interpretations of parallel systems of coded customary law, national laws and regulations, and actual practice (common law). Women's access to decision-making about land and trees, and to land and trees as goods, may require changes in legal codes, their interpretation or their enforcement. This may be as sweeping as a national legislative reform to improve women's legal status as property owners and claimants. It may also be as simple as a change in interpretation of existing national laws in order to facilitate internal reforms of customary law, by ethnic group, to broaden and/or protect and formalize women's rights of ownership and access to land and trees.

5) National policy initiatives can also strengthen the status and power of organizations that represent rural women. While this sometimes results in government or elite manipulation of non-representative organizations, it is also possible to empower authentic, popular women's organizations from grass roots to national levels. Legal legitimacy and widespread official recognition can allow such groups to deal with credit, project administration, access to the factors of agroforestry and social forestry production, directions of research and development, and formulation and enforcement of local codes or project contracts that protect women's rights to land, trees, and their products.

Policy shifts and interventions such as these would provide necessary but not sufficient conditions for women's full participation in agroforestry and social forestry. The process also requires technologies that serve women's interests and effective, responsive organizations for local planning, implementation and enforcement of agroforestry and social forestry programs for the benefit of both men and women.

Technology research and development

The technology development sector of agroforestry and social forestry can help to maintain or improve women's access to land, trees and their products by placing more emphasis on the following design criteria:

1) Choose spatial arrangements that help to maintain or strengthen women's rights and ease of access to places, plants, and products already of use to them.

a) Design for the whole system and the full range of landscape niches, not just cropped plots or titled, fenced farm and community plots (Rocheleau and Hoek 1984).

b) Include technologies for the commons and home gardens, or other landscape niches generally available to and used by women in a given region (this assumes prior survey and "mapping" of women's use of and access to various landscape features).

2) Integrate women's gathered plants and plant products (or acceptable alternatives) into the design of agroforestry and social forestry technologies.

3) Include plants that yield products with potential for profitable processing and/or marketing by women (especially important where tenure is a major constraint to equitable sharing of production decisions or direct production benefits).

4) Disaggregate client groups (except in stable, egalitarian conditions) to allow women to express their own needs, interests, knowledge, expectations, and constraints with respect to problems to be addressed, priorities, technology design and management, and distribution of costs and benefits from proposed technologies (FAO 1982, 1985, Butler-Flora 1984).

5) Involve women clients and women professionals in the entire process of technology research, from problem definition and design and testing of technology prototypes to local adaptation, extension and continued refinement of technologies (Hoskins 1983, Unasylva 1984a).

Participation and rural development

Social forestry and agroforestry action programs often emphasize women's participation as such, with less attention paid to access to resources. However, it is not only participation for its own sake that matters to women clients of such programs. They also care about the concrete performance of the program, the quality of their own participation, and the distribution of the actual costs and benefits.

Among the most important opportunities for direct action are the following:

1) Agroforestry and social forestry programs can use equitable project participation contracts (Hoskins 1983, Bruce and Noronha, this volume) and program regulations to set precedents that can be formalized at district and national level. Policy action on all of the points already discussed could be developed and tested under site- and program-specific conditions, without recourse to slow legislative action or to inflexible national regulations. After careful bottom-up development and testing of new codes, either the codes themselves, or the procedures for developing them could be adopted at the district or national level.

2) Social forestry and agroforestry action programs are fairly new and expanding, which gives them ample scope to train and hire women into skilled and supervisory positions. Women clients often are better able to deal directly with women professionals on matters of access to land, trees and their products, as well as on technology questions relating to women's work. This will vary by ethnic group, class, and education, and will also depend on the communication skills and training of field personnel.

3) Most programs can help women to gain access to agroforestry and social forestry benefits by focusing some of their resources on activities in which women already have some interest, control or rights of access to land, trees or their products. This may involve straying from a single commodity or single activity focus, in order to accommodate existing tenure constraints and/or strict sexual division of tenure.

4) Where women's tree and land tenure and direct access to decision making are unduly constrained, agroforestry and social forestry programs may serve women clients best by working to adapt existing organizations toward a greater voice for women in agroforestry and social forestry decisions controlled by men. Another alternative would be to introduce new types of agroforestry and social forestry activities, sites, and plant types that are not yet defined as men's domains (Fortmann and Rocheleau 1985), and to promote profitable roles for women with as much control by them as possible over the production process and/or the final products.

5) Wherever possible agroforestry and social forestry action programs should work with community organizations that are already able and willing to channel information to and from rural women and to provide a forum for discussion between local women, the community at large and the program. This may imply working with two or more groups instead of just one, or it may mean that a more general purpose community group would be more appropriate than an all men's farmer association.

Conclusion

Much attention is being focused on women's involvement at a time when social forestry and agroforestry are still learning to involve the population at large, and to think in terms of "clients" rather than "targets." Much of the action research and organizational experiments required to find viable rules of tree and land ownership, access and management for women can be nested within broader programs to achieve the same for the whole population. This is not to say that both agendas should be homogenized into a single activity. There is still a need to disaggregate information, decisions and action to ensure reasonable and equitable distribution of land, trees, or their products, and of program costs and benefits to all clients, men and women.

The recommended courses of action for the policy, research and development sectors in agroforestry and social forestry involve ambitious undertakings. However, it is possible to integrate all three into a single approach. Technology research and action research, combined within larger development programs, can feed appropriate experience and a diversity of models back to the policy sector, which in turn can provide more scope for women's access to resources in future agroforestry and social forestry research and action programs.

One of the key elements in the action research programs will be the choice of organizations to represent women and the style of interaction within and between organizations. Adequate information for these decisions will not always be available nor will it necessarily be consistent over large regions. Perhaps the best approach to this issue is to work toward the most effective unit and terms of organization through participatory research (Uphoff 1979, Max-Neef 1982, Lowdermilk and Laitos 1981, Messerschmidt 1983, Woods 1983) in representative areas, allowing women to identify, create or modify the most appropriate units of organization to represent their interests and to mediate their participation in agroforestry, social forestry and land tenure reform activities. It is here that the research frontier beckons strongest and where women's future access to land, trees and their products will be determined.

Acknowledgements

In the course of preparing this paper I have received bibliographic gems as well as ideas and encouragement from several colleagues, including John Raintree, Marilyn Hoskins, Robert Chambers, Louise Fortmann and Carolyn Barnes. Special thanks are due to Paula Williams and John Raintree for their careful review and comments. I am indebted to the authors of the other background papers (L. Fortmann, J. Raintree, J. Riddell, J. Bruce and R. Noronha) and to M. Cernea, R. Wekeza and D. Brokensha and P. Castro for the opportunity to review draft documents for information and or issues bearing on women's access to land and trees.

References

Afonja, S. 1981. Changing modes of production and the sexual division of labour among the Yoruba. *Signs* 7(2): 299-313.

Agarwal, A. and A. Anand, 1982a. *Introducing New Technologies I: Try Asking the Women First.* Report No. 53. Centre for Science and Environment. New Delhi.

Agarwal, A. and A. Anand. 1982b. *Introducing New Technologies II: Try Asking the Women First.* Report No. 54. Centre for Science and Environment. New Delhi.

Ahooja-Patel, K. 1982. Another development with women. *Development Dialogue* 1-2: 17-28.

Arizpe, L. 1982. Women and development in Latin America and the Caribbean: lessons from the seventies and hopes for the future. *Development Dialogue* 1-2: 74-84.

Ayensu, E.S. 1983. The healing plants. *Unasylva* 35(140): 2-6.

Barnes, C. 1984. The Historical Context of the Fuelwood Situation in Kisii District. In C. Barnes, J. Ensminger and P.O' Keefe (eds). *Wood Energy and Households: Perspectives on Rural Kenya.* The Beijer Institute and The Scandinavian Institute of African Studies. Stockholm and Uppsala.

Berreman, G. 1984. From Ghandhi to Chipko: the movement to save the Himalayas. *The Global Reporter* 1(4): 16-18.

Berry, S., J. Geyer, P. Peters, A. Issacman. S. Moore, A. Richard, and M. Watts. 1983. *The Food Crisis and Agrarian Change in Africa.* SSRC/NRC. Washington, D.C.

Bhatt C.P. 1982. Ecosystem of the central Himalayas and Chipko movement, determination of hill people to save their forests. Dashauli Gram Swarajya Sangh, Gopeshwar, Uttar Pradesh.

Bhatt. G.S. 1980. Woman's relation with her habitat in the polyandrous tracts of Uttar Pradesh Himalayas. *Proceedings of the Seminar on the Role of Women in Community Forestry.* Forest Research Institute and Colleges. Dehra Dun, Indian.

Black, N. and A.B. Cottrell (eds). 1981. *Women and World Change: Equity Issues in Development.* Sage Publications Beverly Hills.

Boormans, V. 1984. Technique and women's toil. Paper prepared for a consultation within the Belgian Ministry of Cooperation on Appropriate Technology to Lighten Women's Daily Heavy Task. Brussels.

Boserup, E. 1970. *Women's Role in Economic Development.* George Allen and Unwin Ltd. London.

Boyd, D.J. 1984. Awa women in Papua New Guinea. *Cultural Survival Quarterly* 8(2): 36-37.

Brain, J.L. 1976. Less than second class: women in rural settlement schemes in Tanzania. In N.J. Hafkin and E.G. Bay (eds). *Women in Africa*. Stanford University Press. Stanford, California.

Brokensha, D., B. W. Riley and A. P. Castro 1983. *Fuelwood Use in Rural Kenya: Impacts of Deforestation*. USAID. Washington, D.C.

Brokensha, D. and A.P. Castro. 1984. *Fuelwood, Agroforestry and Natural Resource Management: The Development Significance of Land Tenure and Other Resource Management/Utilization Systems*. USAID Contract No. OTR-0000-0-00-3441-00. Institute for Development Anthropology. Binghamton, New York.

Buch, N. 1980. Role of women in nutrition gardens in social forestry. *Proceedings of the Seminar on the Role of Women in Community Forestry*. Forest Research Institute and Colleges. Dehra Dun, India.

Butler-Flora, G. 1984. Intra-household dynamics in farming systems research: the basis of whole farm monitoring of farming systems research and extension. Kansas State University FSSP. Manhattan, Kansas.

Buvinic M. and Youssef N.H. 1978. Women-headed households: the ignored factor in development planning. International Centre for Research on Women. Washington, D.C.

Cantor, J. 1984. Survey of off-farm land use by 31 households in sample watershed, Kathama Sub-Location, Machakos District, Kenya (Summary). ICRAF. Nairobi.

Caplan, D. 1984. Survey of off-farm land use by households in Kakuyuni watershed, Katangi Sub-Location, Machakos District Kenya. ICRAF. Nairobi.

Cernea, M.M. 1982. Tribal peoples and economics development: human ecologic considerations. World Bank. Washington, D.C.

Cernea, Michael. 1985. Units of social organization sustaining alternative forestry development strategies. Agriculture and Rural Development Department. The World Bank. Washington, D.C.

Chaiken, M. 1985. *Nutritional Surveillance and Intervention Programme. Mbita Division, South Nyanza, Kenya*. UNICEF. Nairobi.

Chakravorty, S. 1980. Nutrition gardens, plantations of quick growing fuel trees-role of women. *Proceedings of the Seminar on the Role of Women in Community Forestry*. Forest Research Institute and Colleges. Dehra Dun, India.

Chambers, Robert. 1983. *Rural Development: Putting the Last First*. Longmans. London.

Chambers, R. 1984. To the hands of the poor: water, trees and land. Discussion Paper No. 14. Ford Foundation. New Delhi.

Chaney E. and M. Lewis. 1980. Planning a family food production program: some alternatives and suggestions for Plan Sierra. Plan Sierra. San Jose de las Matas, Dominican Republic.

Chavangi, N. 1984. Cultural aspects of fuelwood procurement in Kakamega District, Kenya. Working Paper No. 4. Kenya Fuelwood Development Programme/Beijer Institute. Nairobi.

Christophersen, K.A. and F. Weber. 1981. Energy potential from native brushland in Niger: the economic perspective. Report to USAID Office of Energy. (AID/USDA Bioenergy Team; PASA AG/STB-4709-6-79).

Colfer C.J.P. 1981. Women, men and time in the forests of East Kalimantan. *Borneo Research Bulletin* September: 75-85.

Colfer, C.J.P. 1984. An anthropologist among foresters. *Unasylva* 36(146): 33-34.

Cultural Survival Quarterly. 1982. Deforestation, the human costs. *Cultural Survival Quarterly* 6(2).

Cultural Survival Quarterly. 1984a. Introduction. *Cultural Survival Quarterly* 8(2):2-5.

Cultural Survival Quarterly. 1984b. The disinherited. *Cultural Survival Quarterly* 8(2): 24-25.

Dahlberg, F.M. (ed). 1981. *Woman the Gatherer*. Yale University Press. New Haven.

Das Gupta, M. 1984. Microperspectives on the slow rate of urbanization in India: informal security systems and population retention in rural India. Paper presented at the IUSSP Seminar on Micro-Approaches to Demographic Research. September. Canberra.

Deere, C.D. and de Leon, M.L. 1981. Peasant production, proletarianization and the sexual division of labour in the Andes. *Signs* 7(2):338-60.

Endicott, K.L. 1984. The Batek De' of Malaysia. *Cultural Survival Quarterly*. 8(2):6-8.

FAO. 1979. The legal status of rural women: limitations on the economic participation of women in rural development. FAO Economic and Social Development Paper 9. FAO. Rome.

FAO. 1982. *Guidelines: Women in Land and Water Development.* Land and Water Development Division. FAO. Rome.

FAO (ESHE). 1983a. *Follow-up to WCARRD: The Role of Women in Agricultural Production.* Expert Consultation on Women in Food Production. (ESH:WIFP/83/11) 7-14 December. Rome.

FAO (ESHE). 1983b. The need for improved agricultural extension services for women engaged in agriculture. Expert Consultation on Women in Food Production (ESH;WIFP/83/3-Rev. 1). Rome.

FAO. 1985. "Tree Growing by Rural People." Review Draft. Forestry Policy and Planning Service. FAO. Rome.

Flora, C.B. and B. Santos. 1984. *Women in a Dominican Republic Farming System.* FSSP/Kansas State University. Manhattan, Kansas.

Flores-Paitan, S. 1985. Informe solue el papel de Umari en los sistemas de produccion agroforestal en fincas de la pollacion indigena y los mestizos en la zona de Iquites. University of Iquitos. Iquitos, Peru.

Foley, G. and G. Barnard. 1983. "Social Forestry--A Reappraisal." (Draft). Earthscan/I.I.E.D. London and Washington, D.C.

Fortmann, L. 1981. The plight of the invisible farmer: The effect of national agricultural policy on women in Africa. In *Women and Technology Change in Developing Countries*. R. Dauber and M. Cain (eds). AAS Selected Symposium 53. AAAS. Washington D.C.

Fortmann, L. 1984. Economic status and women's participation in agriculture: A Botswana case study. *Rural Sociology* 49(3): 452-464.

Fortmann, L. 1985. The tree tenure factor in agroforestry,with particular reference to Africa. *Agroforestry Systems* 2: 229-251.

Fortmann, L. and D. Rocheleau. 1984. Why agroforestry needs women: four myths and a case study. *Unasylva* 36:146.

Fortmann, L. and D. Rocheleau. 1985. Women and agroforestry: four myths and three case studies. *Agroforestry Systems* 2:253-272.

Gordon, Elizabeth. 1981. An analysis of the impact of labour migration on the lives of women in Lesotho. *The Journal of Development Studies* 17(3):65.

Harman, I.M. 1984. Women and cooperative labour. *Cultural Survival Quarterly* 8(2):38-40.

Haugerud, A. 1983. The consequences of land tenure reform among smallholders in the Kenya Highlands. *Rural Africana* 15/16: 65-90.

Hoskins, Marilyn. 1979. *Women in Forestry for Local Community Development: A Programming Guide*. Office of Women in Development AID/Otr-147-79-83 Washington D.C.

Hoskins, M.W. 1980. Forestry and rural women. *Proceedings of the Seminar on the Role of Women in Community Forestry*. Forest Research Institute and Colleges. Dehra Dun, India.

Hoskins M. 1982 Social forestry in West Africa: myths and realities. Paper presented at the annual meeting of the American Association for the Advancement of Science. Washington, D.C.

Hoskins, M. 1983. Rural women, forest outputs and forestry projects. FAO. Rome.

Hunt, Diana. 1984. *The Impending Crisis in Kenya: The Case for LandReform*. Gower. Hampshire.

Hunink, R.B.M. and J.W. Stoffers. 1984. *Mixed and Forest Gardens on Central Java: An Analysis of Socio-Economic Factors Influencing the Choice Between Different Types of Land Use*. Department of Geography of Developing Countries, Utrecht University. Utrecht.

Huxley, P.A., D. Rocheleau and P.J. Wood. 1985. *Farming Systems and Agroforestry Research in Northern Zambia: Phase I Report -- Diagnosis of Land Use Problems and Research Indications*. ICRAF. Nairobi.

ICES. 1985. Beyond cultural diversity, women face some problems. IFDA Dossier 46: 27-29. International Centre for Ethnic Studies. Nairobi.

ICRAF. 1983. Guidelines for agroforestry diagnosis and design. ICRAF. Working Paper No. 6. ICRAF. Nairobi.

ICRAF. 1983. Resources for agroforestry diagnosis and design. Working Paper No. 7. ICRAF. Nairobi.

IDRC Reports. 1984. Commentary--observing the overlooked: New ways of gathering information for research on women. *The IDRC Reports* 13(2): 22-23.

Jain, D. 1984. In poor families women's income is the lifeline. *Ceres* 17(4): 35-38.

Jain, S. 1984. Standing up for the trees: women's role in the Chipko Movement. *Unasylva* 36(146):12-20.

Jiggins, J. (n.d.) Female-headed households: Mpika sample, Northern Province. Occasional Seminar 3, RDSB. University of Zambia.

Joshi, G. 1982. *Men Propose, Women Oppose the Destruction of Forests*. Information Service on Science and Society-Related Issues. Centre for Science and Environment. New Delhi.

Kelly, M.P.F. 1981. The sexual division of labour, development, and women's status. *Current Anthropology* 22(4): 414-419.

Kerven, C. 1984. The family farm and outmigration: some issues for FSR in Africa. Paper presented at the ARPT/CIMMYT Workshop on the Role of Rural Sociology in Farming Systems Research and Technology Generation. Lusaka, Zambia.

Kishindo, P. 1984. Agricultural development in a matilineal system. Paper presented to the ARPT/CIMMYT Workshop on the Role of Rural Sociology in Farming Systems Research and Technology Generation. Lusaka, Zambia.

Ki-Zerbo J. 1981. Women and the energy crisis in the Sahel. *Unasylva* 33(133): 5-10.

Lagemann, J. 1977. *Traditional African Farming Systems in Eastern Nigeria*. Westforum Verlag. Munick.

Lee, R.B. 1979. *The Kung San: Men, Women, and Work in a Foraging Society*. Cambridge University Press. New York.

Livingstone, R. and R. Zamora. 1983. Medicine trees of the tropics. *Unasylva* 35(140): 7-10.

Lowdermilk, M. and W.R. Laitos. 1981. Towards a participatory strategy for integrated rural development. *Rural Sociology* 46(4): 688-702.

March K. and R. Taqqu. 1982. Women's informal associations and the organizational capacity for development. Cornell University Rural Development Committee. Ithaca.

Max-Neef, M.A. 1982. *From the Outside Looking In: Experiences in "Barefoot Economics"*. Dag HammarsKjold Foundation. Uppsala.

Maybury-Lewis, P. 1984. One step forward-two steps back. *Cultural Survival Quarterly* 8(2):12-14.

Mayer, A.E. 1984. Law and women in the Middle East. *Cultural Survival Quarterly* 8(2):49-50.

Mazumdar, V. 1982. Another development with women: a view from Asia. *Development Dialogue* 1-2: 65-73.

Mazumdar, V. 1984. With progress for some: women and development in India. *The IDRC Reports* 13(2): 4-6.

Messerschmidt, A.D. 1983. *Gaun Sallah: The "Village Dialogue" Method for Local Planning in Nepal.* Ministry of Forest and Soil Conservation, RCUP. Kathmandu, Nepal.

Mickelwait, D.R., M.A. Riegelman., and C.F. Sweet. 1976. *Women in Rural Development: A Survey of the Roles of Women in Ghana, Lesotho, Kenya, Nigeria, Bolivia, Paraguay and Peru.* Westview Press. Boulder, Colorado.

Mishra A. and S. Tripathi. 1978. *Chipko Movement: Uttarakhand Women's Bid to Save Forest Wealth.* People's Action. New Delhi.

Muchena, O.N. 1984. Strategies for women's organizations in developing countries. Paper presented at Second International Interdisciplinary Congress on Women: Strategies for Empowerment. Gronigen, Netherlands.

Muntemba, Shimwaayi. 1982. Women as food producers and suppliers in the twentieth century: the case of Zambia. *Development Dialogue* 1-2: 29-50.

Nambombe, V. 1984. Diagnosis of agroforestry potentials in a pastoral system: Kiboko Group Ranch, Kajiado District, Kenya. On-the-job Training Programme Final Seminar Report. ICRAF. Nairobi.

Naqvi, S. and K.M. Gupta. 1980. Social forestry and the role of hill women--some pre-conditions. *Proceedings of the Seminar on the Role of Women in Community Forestry.* Forest Research Institute and Colleges. Dehra Dun, India.

Ninez, A. 1985. Introduction: household gardens and small-scale food production. In V. Ninez (Guest Editor). *Special Issue on Household-Level Food Production.* Food and Nutrition Bulletin 7(3): 1-5.

Nowak, B.S. 1984. Can the partnership last? *Cultural Survival Quarterly* 8(2): 9-11.

Obi S.N.C. 1983. *The Ibo Law of Property*. Buttersworth. London.

Okoth-Ogendo, H.W.O. 1981. The law in relation to land practices in Kenya. In *Proceedings of the Kenyan National Seminar of Agroforestry*. ICRAF. Nairobi

Okeyo-Pala, A. 1979. Women in the household economy: managing multiple roles. *Studies in Family Planning* 10(11/12).

Okeyo-Pala, A. 1980. Daughters of the lakes and rivers: colonization and the land rights of Luo women. In M. Etienne and E. Leacock (eds). *Women and Colonization: Anthropological Perspectives*. Praeger Publishers. New York.

Onger-Hosgor, T. 1983. The effects on women of land tenure changes and agrarian reform. Expert Consultation on Women in Food production. (ESH:WIFP/83/12) 7-14 December. FAO. Rome.

Palmer, I. 1978. Integration of women in agrarian reform and rural development: an analysis of the situation in Asia and the Far East region. FAO. Rome.

Papanek, H. 1981. The Differential Impact of Programs and Policies on Women in Development. In R. Dauber and M. Cain (eds). *Women and Technological Change in Developing Countries*. AAAS Selected Symposium 53. Westview Press. Boulder, Colorado.

Patil, J. 1980. Role of tribal women in social forestry and rural developments. *Proceedings of the Seminar on the Role of Women in Community Forestry* Forest Research Institute and Colleges. Dehra Dun, India.

Peacock, N. 1984. The Mbuti of Northwest Zaire. *Cultural Survival Quarterly* 8(2): 15-17.

Peters, P. 1984. Household management in Botswana: cattle, crops, and wage labour. Paper prepared for the Joint Rockefeller and Ford Foundation Conference on Intra-Household Processes and Farming Systems Analysis. Bellagio, Italy.

Raintree, J.B., D. Rocheleau, P. Huxley, P. Wood and F. Torres. 1985. Draft report on the joint ICAR/ICRAF diagnostic and design exercise at the Bhaintan watershed in the outer Himalayas of Uttar Pradesh. ICRAF. Nairobi.

Raintree, J.B. and K. Warner. 1985. Agroforestry pathways for the integral development of shifting cultivation. IX World Forestry Congress. Mexico City.

Rocheleau, D. and A. van den Hoek. 1984. The application of ecosystems and landscape analysis in agroforestry diagnosis and design: a case study from Kathama Sublocation, Machakos District Kenya. ICRAF Working Paper No. 11. ICRAF. Nairobi.

Rocheleau, D. 1984a. An ecological analysis of soil and water conservation in hillslope farming systems: Plan Sierra, Dominican Republic. Unpublished Ph.D dissertation. Dept. of Geography, University of Florida. Gainesville.

Rocheleau, D. 1984b. The application of the D&D methodology at varying acales of analysis. In E. Zulberti (ed). *Agroforestry Training Materials*. ICRAF. Nairobi.

Rocheleau, D. 1985. Criteria for re-appraisal and re-design: intra-household and between-household aspects of FSRE in three Kenyan agroforestry projects. In Flora and Tomacek (eds). *Selected Proceedings; Annual Symposium on Farming Systems Research and Extension Oct. 7-18, 1984*. Kansas State University. Manhattan, Kansas.

Roy, S. 1980. Case study on fuelwood collection and problems of rural women. *Proceedings of the Seminar on the Role of Women in Community Forestry* Forest Research Institute and Colleges. Dehra Dun, India.

Russell, M. 1984. Clearer focus: a plea for more small-scale face-to-face interviewing in contemporary African social research. Paper presented to the ARPT/CIMMYT Workshop on the Role of Rural Sociology in Farming Systems Research and Technology Generation. Lusaka, Zambia.

Ruthenberg, H. 1976. *Farming Systems in the Tropics*. Clarenden Press. Oxford.

Scott G. 1980. Forestry projects and women. The World Bank. Washington, D.C.

Seidman, A. 1981. Women and development of "underdevelopment": the African change. In R. Dauber and M. Cain. (eds). *Women and Technological Change Developing Countries*. AAAS Selected Symposium 53. Westview Press. Boulder, Colorado.

Sharma, R. 1981. Greening the countryside. Information Service on Science and Society-Related Issues. Centre for Science and Environment. New Delhi.

Skutsch, M.M. 1983. Why people don't plant trees. Discussion Paper D-73P. Energy in Developing Countries Series. Resources for the Future. Washington, D.C.

Sommers, P. 1978. Traditional home gardens of selected Philippines households and their potential for improving human nutrition. M.Sc. thesis. University of the Philippines. Los Banos.

Stoler, A. 1978. Garden use and household economy in rural Java. *Bulletin of Indonesian Economic Studies* 14(2): 85-101.

Stollen, K.A. 1983. Peasants and agricultural change in northern Zambia. Occasional Paper 4. Department of Agricultural Economics. Agricultural University. AS, Norway.

Surin, V. and T. Bhaduri. 1980. Forest produce and forest dwellers. *Proceedings of the Seminar on the Role of Women in Community Forestry.* Forest Research Institute and Colleges. Dehra Dun, India.

Tanner, R. 1960. Land rights on the Tanganyikan coast. *African Studies* XIX: 14-25.

Thomson, J. 1983. *Participation, Local Organization, Land and Tree Tenure: Future Directions for Sahelian Forestry.* Club du Sahel/OECD. Paris.

Tinker, I. 1976a. *The Real Energy Crisis: Women's Time.* EPOC. Washington, D.C.

Tinker, I. 1976b. The adverse impact of development on women. In I. Tinker and M.B. Bramsen (eds). *Women and World Development.* Praeger Publishers. New York.

Tinker, Irene. 1981. New technologies for food-related activities: An equity strategy. In R. Dauber and M. L. Cain (eds). *Women and Technological Change in Developing Countries.* AAAS Selected Symposium 53. AAAS Washington, D.C.

Unasylva. 1984a. Women in forestry. *Unasylva* 36(146).

Unasylva. 1984b. Effects of scarcity upon women. *Unasylva* 36(146):30.

Unasylva. 1984c. Forestry projects for women only? *Unasylva* 36(146):30.

Uphoff, N.T., J. Cohen, and A. Goldsmith. 1979. Feasibility and application of rural development participation: a state of the art paper. Rural Development Committee. Cornell University. Ithaca, New York.

Vayda, A.P., C.J.P. Colfer, and M. Brotokusumo. 1980. Interactions between people and forests in E. Kalimantan. *Impact of Science on Society* 30(3). Reprint No. 13 East-West Center Environment and Policy Institute. Honolulu.

Wake, J.L. 1984. The cost of learning by doing: Effect on technology adoption. Unpublished M.S. thesis. University of Florida. Gainesville.

Weber, F. and M. Hoskins. 1983. *Agroforestry in the Sahel*. Department of Sociology. Virginia Polytechnic Institute and State University. Blacksburg, Virginia.

Whalen, I. 1984. Land tenure and technological innovations in the Ethiopian highlands. Paper prepared for the ARPT/CIMMYT Workshop on the Role of Anthropologists and Rural Sociologists in Farming Systems Research and Extension. Lusaka, Zambia.

Whyte, W.F. 1982. Social inventions for solving human problems. *American Sociological Review* 47.

Wiff, M. 1984. Honduras: Women make a start in agroforestry. *Unasylva* 36(146): 21-26.

Wijngaarden, J. van 1983a. Agricultural self help groups and their potential role in agroforestry. Draft Report. Wageningen Agricultural University. Wageningen, The Netherlands.

Wijngaarden, J. van 1983b. Patterns of fuel gathering and use. M.Sc. Thesis. Department of Forestry Management, Wageningen Agricultural University. Wageningen, The Netherlands.

Williams, P.J. 1982. Women and forest resources. In M. Stock, J. Force and D. Ehrenreich (eds). *Women in Natural Resources: An International Perspective*. Proceedings of a conference for men and women held at the University of Idaho.

Williams, P.J. 1983. The social organization of firewood procurement and use in Africa: A cross-cultural study of the division of labour by sex. Ph.D. Dissertation. University Microfilms. Ann Arbor, Michigan.

Williams, P.J. 1984. La periode de soudure. Teport on human use of forest resources in Burkina-Faso. Institute of Current World Affairs. Hanover, New Hamphire.

Williams, P.J. 1985. Women and forestry. Invited Special Paper. IX World Forestry Congress. Mexico City.

Woods, B. 1983. *Altering the Present Paradigm: A Different Path to Sustainable Development in the Rural Sector*. The World Bank. New York.

Young, W.C. 1984. Cultural change and women's work: The sedentarization of the Rashiidy Bedouin in the Sudan. *Cultural Survival Quarterly* 8(2) 28-29.

Land Tenure Issues in the Forestry and Agroforestry Project Contexts

John W. Bruce and Raymond Noronha

Land Tenure Center, University of Wisconsin-Madison;
Land Tenure Consultant, IBRD

Introduction

This paper has six sections. In the first, it suggests several ways in which ignoring or poorly handling land tenure issues can contribute to project failure or shortfall. In the second, it suggests a series of inquiries concerning traditional tenure rights which must be a part of the project planning process if tenure issues are to be recognized and understood. In the third, it examines the nature of tenure issues which generally arise in social/community forestry project contexts in which trees are planted on community land. In the fourth, it examines the rather different tenure issues which arise in agroforestry projects in which trees are planted on the holdings of individuals or households. In the fifth, it discusses several options in response to tenure issues identified in project planning, and the legal and administrative tools which may be available to project planners and managers. Finally, the sixth section attempts to suggest areas where research is especially needed.

1. Land tenure and project failure/shortfall

Projects come to grief over land tenure because they are designed on incorrect assumptions--explicit or implicit--about land tenure in the project area. At the project identification stage a given land tenure situation is pregnant with both opportunities and constraints. A project develops land tenure "problems" only if it is unrealistically designed, with constraints ignored and opportunities overlooked. We should be clear at the outset that in talking of land tenure problems in projects we are dealing with poor project design, usually project design which focuses too heavily on new practices or techniques to be adopted and too casually on socioeconomic factors affecting adoption. The task of realistic assessment of the opportunities and constraints posed by a land tenure situation is a part of the more general task of adequate assessment of the socioeconomic environment of the project, but a particularly important part because it deals directly with land, the resource on which all agricultural development is based.

This section of the paper seeks to identify some of the common causes of project failure or shortfall due to neglect of land tenure issues. Four categories have been identified. The first two involve failure to frame a project so as to be consistent with land tenure realities, the third concerns the failure of new tenure arrangements created for project beneficiaries, and the fourth concerns problematic land tenure situations created by project activities.

Neglect of constraints on farmer behavior

Inadequate understanding of existing land tenure arrangements can cause project planners to neglect social and institutional constraints which prevent farmers from responding to opportunities posed by the project. This may involve either a simple lack of knowledge of constraints posed by customary norms or an underestimation of their durability and persistence. Sometimes a customary constraint is ignored due to erroneous reliance on a formal legal position. A common mistake is to assume a greater exclusivity of use and management by primary users than is the case. Frequently land is used serially for different purposes in different seasons by different users, and a major intensification of use such as tree planting can exclude other uses by other individuals, the community or other groups (see Hammer 1983).

Miscalculation of farmer incentives

Inadequate understanding of existing tenure arrangements can cause project planners to misjudge the incentive effect of project-created opportunities, with the result that farmers reject those opportunities. In some cases there is a failure to understand the impact on farmers' incentives of the physical configuration of holdings created by the tenure system, as in systems characterized by fragmented holdings or by maldistributed and economically marginal holdings (Ampuero 1979). Alternatively, there may be a failure to appreciate the effect of tenure situations where investment costs must be borne by one party but the benefits of the investment shared with others (Bruce 1983) In such cases the rules of the traditional system do not forbid the investment, but a set of circumstances arising under them can affect farmers' incentives and decisions concerning investments.

Inadequate framing of new tenure arrangements

There are projects in which the need for new tenure arrangements is appreciated, but those arrangements are poorly conceived or implemented. This is most common in settlement projects, including settlement in irrigation projects. Because agroforestry projects most often take place on existing holdings rather than "schemes" created for the purpose, such difficulties may not often arise. But whenever tree crops are being introduced on a scheme basis, either exclusively or as part of a mixed cropping system, considerable care must accompany the framing of tenure arrangements. Insufficient security of tenure for participants can lead to abandonment of the seedlings.[1]

Tenure problems created by project-induced change

Projects themselves sometimes create new situations which give rise to land tenure problems. Some of these affect quite directly the success of the project in question; others, while they do not always hinder the achievement of project objectives, may interfere with important land tenure and other public policies.

An example is the problem of beneficiary displacement. Those cultivators who are the intended beneficiaries of the project are pushed off their land by others, attracted to the land by the project. Beneficiary displacement usually takes place due to a combination of project-generated appreciation in value of land that was previously marginal, on one hand, and on the other ambiguous or insecure title on the part of the intended beneficiaries. The beneficiaries' use of the marginal land appears uncontested until its value increases; then other, more powerful claimants come forward. The result may be beneficiary displacement or a stalemate preventing *anyone* from benefiting from the project. The other claimants may be local individuals or groups, or members of national elite. In Cebu Province in the Philippines, for instance, farmers oppose government efforts at tree planting, realizing that it may revive the interest of absentee claimants to the lands they farm (Suganob 1980). Particularly in the case of elites, the displacement may be pre-emptive, involving land-grabbing prior to the project activities where the project will provide an attractive economic opportunity to landholders in the project area.[2]

Displacement may involve others than the intended beneficiaries, and can arise not only out of an increase in land value but out of a change in land use. Such a change in use can render impossible the continued overlapping or serial use of land by different individuals or groups. In Swaziland, for example, there has been resistance to new crops and fencing of holdings where the new crops interfere with the turning of stock onto the farmland in the winter months (LTC 1985). Attempts to increase security of tenure for one right-holder means eliminating or reducing rights of others, sometimes in ways not clearly anticipated (Eckholm 1967).

Given that problems such as these do arise, what are the kinds of inquiries that need to be made at the project design stage in order to ensure early awareness of problems and the opportunity to avoid or resolve them?

2. Analytical basis for interpreting land tenure and land use systems

The determination of who is in possession of the land, that is, who has the right to use it and who is actually using it, is necessary since it would provide answers to whether there is land available for the project, and also influence project design and scheduling. The answers can only be obtained through investigations "on the ground," even in countries that have cadastral surveys and provisions for title registration. For, as explained in greater detail below, actual patterns of land use can differ significantly from apparent, formal-legal patterns. It should also be remembered that cadastral surveys are carried out at intervals of ten years or more, and that title registers often do not contain notations about sharecropping arrangements, or traditional allocations of land (for instance, to wives and sons).

The basic question is who has an interest in land?[3] An answer to this question can be subdivided into several interrelated enquiries: Is the land being used? Who is the "owner" or has the right to allocate the land? Are distinctions drawn between types of land? Who is the allottee of the land or who uses it? What are the rights of the allottee? How is the land actually used? Are distinctions drawn between different types of crops?

Existing rights to the land

Is the land being used?

One of the first questions to be answered at the stage of project identification and preparation, particularly in settlement and plantation projects, is whether there is land available for the project. There is a difference between land without title and land that is not being used. Land may not be used for several reasons related to the technology employed, the nature of the soils, seasonal climatic variations, and the type of user. For example, land may appear unused because shifting cultivation is practiced by the group and there are no boundary marks (except in the minds of members of the group).[4] This is the case with the majority of cultivation in Papua New Guinea, parts of the Philippines, Malaysia, Indonesia, Africa and Latin America. Again, the land may appear unused to project designers (government staff, local or international consultants) who arrive during the off season, the fallow period, or when the pastures and water have been exhausted and the transhumants have moved to other lands. Or, again, the land may be used in common and may, therefore, have no apparent individual use. Or, finally, even though the land is not presently being used, transhumant groups may have a traditional right to pass, and re-pass, over defined routes on the land.

Who is the "owner"? Who has the right to allocate land?

Statutory classifications of "ownership" can be quite misleading. Thus, for example, in Thailand one-half of the country's total area is classified as reserved forest, "owned" by the government. And yet, ten percent of the total population have occupied some of these areas "illegally" for centuries. Project designers view with some trepidation the fact that the "owner" is a group--a tribe, for instance, comprising all members dead, living, and yet to be born (as is recounted of most African tenurial systems, or those of Papua New Guinea, the Pacific and Amazonia).

The term "owner" may cover a *tribe*, as for instance the Jivaro (Brazil), some Liberian tribes, and in Barbuda (the Caribbean); a *lineage* (a group of persons claiming descent from a common ancestor)--as among the Anlo Ewe (Ghana), Kalingas (Philippines), Luo and Kikuyu (Kenya), Ndebele (Zimbabwe), Kongo (C.A.R.), Basseri and Tuareg (Iran and Niger), the Aymara (Peru), Tiv (Nigeria), Sere (Senegal), in Fiji, the Solomon Islands, and Papua New Guinea; an *extended family* (India, Nepal); a *nuclear family* (Afar -Ethiopia, and under Islamic law); an *individual person* (the Caribbean, except Barbuda); and a *legal person* such as a trust--*wakfs* (religious trusts under Islamic law), *guthi* (religious trust) lands in Nepal; an *idol* (India); a *village* in Indonesia or Mexico, among the Iban of Malaysia, Samoa, Tigre (Ethiopia); or the *State*.

For the project designer, searching for the "owner" of the land in these circumstances, or trying to incorporate the "owner" into a project's organizational structure can be a fruitless endeavour, especially where there are no cadastral surveys. More so when the "owner" is a mere figurehead. What is important for the project designer is to find the person, or persons, who have the right to allocate land. In practice, this usually comprises a smaller, locally based group, that has the right to allocate and the right to use the land. It is essential to find out which group or person (as, for instance, the "earth priest" in northern Burkina Faso) has the right to allocate lands since this determination would influence project organization, the choice of decision-making levels, and the persons to whom project communications may have to be directed to achieve project goals.

There could, in fact, be several decision-making groups: the tribe as notional owner; the lineage or sublineage as the allocator of land; and, at the lowest level, within the extended or nuclear family, the eldest male (or female) who further allocates land among members of his family, e.g to his wives, as among the Luo (Okeyo-Pala 1978); or sons, as among the Valley Tonga of Zambia or in Papua New Guinea.

At times, allotments at the lowest level are made to defeat the provisions of existing statutory law (as, for example, in Guyana, Haiti, Brazil, Papua New Guinea Zimbabwe, Turkey). For instance, in both Turkey and Brazil, statutory law requires an equal division of land among all heirs. In Turkey, traditional practices either deny the daughter a share by claiming that it has been paid as dowry, or allow the daughter a share only equal to half that of a male, which is what traditional Islamic law provides (Coulsen 1971). In Brazil, arrangements are usually arrived at between elderly parents (or parents-in-law) and the principal heir, under which the parents gradually sell their property to the principal heir or to the child who cares for them in their old age. Though allocations to wives (in Ghana, Upper Volta, Kenya, Lesotho), or to children (in Papua New Guinea and Brazil) have been termed an allotment, they have to be investigated since such possession affects project design, and could affect an assessment of project targets. For instance, it may be difficult to convince the allottee to purchase inputs to be provided under the project.

Distinctions between types of land allotted

Nearly every group makes careful distinctions between what types of land can be allotted to, and enjoyed by, an individual (or a family, or a sublineage) to the exclusion of others, and what types of lands are open to common use by all the members of the group and, at times, even outsiders. These distinctions are mainly based on two factors:

first, the scarcity, or economic value of the land; and second, the technology employed.

Traditional groups are well aware of the physical and economic attributes of the land, and tenure is determined accordingly. For instance among the Kapauku of New Guinea (now Irian Jaya), land was divided into fourteen categories based on six contrasting components (for example, distribution of vegetation, cultivable potential, soil qualities). Another common distinction is between upland and lowland, as among the Toucouleur (Senegal), Valley Tonga (Zambia), and the Malays (Boutilier 1963, Wong 1975). Where the lowlands are fertile and the uplands not so, tenure rules may be more strictly enforced in the lowlands than the uplands. Among the Mandaya of the Philippines, the upland branch were mainly swidden cultivators of rice, whereas the lowland group were sedentary cultivators of abaca (*Musa textilis*, Manila hemp). Tenure rights were longer among the lowland group and strictly enforced; the upland group moved each year to a new hillside area, retaining rights over only the previous season's land (Yengoyan 1964).

There is a third type of land--the kampong (Malay), the kandyan garden (Sri Lanka), homestead land among the Meru (Kenya), in Zanzibar and in Lesotho--which regularly seems to be owned by individual families. This little area around the household, planted with a few trees, spices, and vegetables, is usually excluded from other tenurial rules. In Sri Lanka they are, even today, exempt from land ceiling laws.

Why is it necessary for the project designer to look into the types of lands? First, because the more valuable they are, the greater the likelihood that the group that uses them will be well-defined and small; therefore, they can be communicated with more easily for project purposes. Second, because traditional groups are more aware of soil potential than others and form a useful source of information for the project designer, information which might prove valuable with regard to the type of crops to be introduced under the project. Third, the type of land and its use is often related to traditional systems of division of labor and may affect both relations between the sexes and the ability to introduce other types of crops or cropping systems. Fourth, the current use of different types of land would also affect assessments of labor availability for the project (see below). Fifth, to enable the project designer to assesss alternative uses to which the land is being, or can be, put.

The scarcity value of land and the growing tendency toward individualization of the tenurial system is assisted not only by the attitudes of governments (and statutory laws encouraging this trend) but, more importantly, by population growth, and the search for cultivable lands to feed the population. There are numerous examples of the influence of population growth on land tenures. For example, population growth in Latin America has tended to make tenurial systems more rigid (Barraclough and Domike 1966). Among the Nupe of Nigeria, increasing population pressure led to a complex system of landownership and transfer which amounted, in effect, to outright transfers for cash (Nadel 1942). Among the Orokaiva of Papua New Guinea, the closer identification of individuals with small tracts of land appeared to be similarly related to high population density (Rimoldi 1966). Again, among the Nyakyusa of Tanzania, increasing population gradually converted shifting cultivation with intergenerational rotation of plots into inheritance of land within families (Wilson 1963). It is increasingly difficult for groups to adjust to the need for land by migrations--such as those by the Mossi of Ghana to the Ivory Coast. There is increasingly less land available for cultivation and settled agriculturists have moved into marginal areas, reducing grazing lands, felling the tree cover, heightening conflicts with herders, and increasing soil erosion hazard. This has happened in Nepal, in the Sudan and in Niger (Macfarlane 1976).

Land use and tenurial patterns are also related to the type of technology employed. Where the technology is simple--a hoe or dibble stick--it makes little sense for individual tenures to exist where the land cannot be cultivated or grazed for more than one season. Populations that practice shifting cultivation usually need more extensive land areas and "own" land in common although it is a smaller unit within the group that allocates land. What is important for the project designer in these cases is not to dismiss the technology out of hand as inefficient but, rather, to enquire into the reason why the technology is employed. The technology cannot be attributed only to lack of contact with a more "civilized" world, but could also be an ecological adaptation to the land and its potential, as in the Amazon and Congo (see for example Renard and Renard 1980, Carneiro 1968, Meggers 1971). The project designer may thus discover that proposed introductions of project technology, or crops, will result in depletion of soil resources and that the innovations will cause greater harm than good in the long term. Finally, since the type of technology used is related to the traditional division of labor, the project designer must take it into account in assessing the probable impact of the introduction of project inputs on this division of labor, on traditional rights, and on the availability of labor for the project.

Do traditional tenurial systems distinguish between types of crops cultivated?

The discussion here will be confined mainly to two types of distinctions: that between subsistence and cash crops; and between seasonal and long-term crops. Seasonal cash crops give rise to greater rigidity and individualization of tenures; there is also a difference in the traditional division of labor. For instance, the cultivation of abaca by lowland Mandaya necessitated longer tenurial rights in the land. Because there was also a need for land for cultivation of food crops, increased land pressures raised the value of land. From a system in which both sons and daughters inherited land, the lowland Mandaya moved to a unilineal inheritance system in favor of sons. Similar changes are taking place in Zambia. Even greater tenurial changes take place with the introduction of semi-permanent (or long-term) crops. Examples are to be found in Liberia, Cameroon and the Ivory Coast. In this category fall economic trees (oil palm, coffee, cocoa, durian and other fruit trees).

The first test to be applied with regard to trees is whether the traditional tenure rules treat them as being attached to the land. If they are deemed to be attached to the land, then they pass to the person in possession of the land--as, for instance, in India and in parts of Kenya. On the other hand, in many parts of the world, trees are not treated as being part of the land, in which case they belong to the person who planted them and are treated as individual property. In practice, problems arise with regard to the planting of trees since it could amount to a claim of ownership of the land. For example, in Lesotho, Mauritania, Ghana, Zanzibar, the planting of trees on arable lands, particularly by tenants is discouraged since this would amount to a claim of ownership of the land. The other problem that exists when traditional rules separate possession of land from ownership of trees is that this separation creates a right for the planter of the trees to enter on the land, a source of constant friction. Further, as in Malaysia, kampong land rights can always be resumed so long as the fruit trees survive--this adds to the problem of determining when the former owner may return and claim possession of the land. Where mortgages are permitted--as they are, for example, in Ghana, Malaysia, and Kenya--the land may be mortgaged to one individual while the trees are pledged to another (for instance, oil palm in Nigeria). The mortgagee of the land could then plant trees (for instance, kola nuts in Zanzibar) in addition to those already on the land, which could add further complications to the tenure and rights both to the land and to the trees.

Different rights, however, exist when the trees are to be used for timber or fuelwood. In this case, the general test appears to be whether the trees are growing wild, or whether they were planted. In the former case, the general rule is (with some exceptions as, for instance, among the Twa of Rwanda) that members of the lineage (or locally based group) have equal rights of access to forest lands. Where, however, trees are planted, they belong to the person who has planted them.

The main reasons for the difference in tenurial arrangements between annual crops and trees appear to lie in the fact that trees take longer to mature (and, therefore, access to them should be maintained), that they require a great deal of labor (for instance, the oil palm, cocoa or citrus), and that they have an economic or scarcity value.[5]

In light of the above discussion it should be realized that it is important for the project designer to examine what crops are cultivated in the project area, who cultivates them, what the division of labor is, and how the produce is appropriated. This examination is necessary because the introduction of a crop could change the tenurial system, lead to resistance to the introduction, and alter traditional rights. Examples of these changes can be found in the Ivory Coast, Ghana and Cameroon. Further, since the introduction could also affect the availability of labor, project designers should assess whether double-cropping should be introduced, and whether it is at all possible to obtain labor for the second crop. For instance, Meek (1949) points out that "In Nigeria an attempt to introduce green manuring had to be abandoned in a certain area when it was found that, under the local system of tenure, women were entitled to grow their minor crops among the major crops of men."

Possession of the land

Who has physical possession of the land?

It is extremely important for the project designer during the preparation stage to make on-the-ground enquiries as to who is actually in possession of the land. Is it the "owner," a tenant, a sharecropper, a wife, a son? Depending on the answers to these enquiries, project components such as extension of credit would have to be designed differently. These enquiries can therefore be divided into two subsidiary questions: How is the land being used (in other words, what is the division of labor)? Are there any traditional labor-sharing arrangements in land use?

Under most traditional systems of land use there is a division of labor in agriculture by sex: the men do the heavy work (clearing, ploughing), women do the planting, weeding, harvesting, threshing --possibly the only exception in agriculture is Haiti where the men do an estimated two-thirds of the work (UNDP 1980, Zeiderstein 1979). This traditional division extends both to the types of lands and crops cultivated. For instance, in parts of Sierra Leone and the Gambia, upland rice is grown as a subsistence crop by women; swamp rice is grown mainly as a crop for sale by men, with women assisting in the seeding and weeding. Men control the proceeds of swamp rice sales; women, therefore, complain that their further involvement in extending swamp rice cultivation cuts into their incomes. As was observed in the case of a rural development project in Sierra Leone, it was extremely difficult to extend swamp rice cultivation, since it required substantial change in the traditional farming system. At the same time, swamp rice cultivation was more influenced by price changes than upland subsistence rice crops where production decisions were based to a greater extent on cultural traditions than market prices. In Liberia it is the reverse: swamp rice is cultivated by women; upland rice, by men. There is also the question of who should obtain credit under a project--if women are cultivators, then credit has to be directed to them. In effect, where traditional practices are affected, the project should be designed to develop measures compensating those affected, and should include carefully prepared communications and extension components.

The traditional division of labor also influences labor availability for project works--at times crucial to project success. For example, under the *gandu* system in northern Nigeria, wives and sons first work on their family plots, followed by work on their own plots. In a livestock project in Senegal, firebreak construction had to be abandoned because of the difficulty of getting laborers. In these circumstances, would labor be available for, say, the cultivation of cash crops, or for the construction of fences?

Labor availability can only be assessed in the light of traditional practices--the division of labor, patterns of migration--not on the basis of an abstract assessment of census statistics of available manpower (statistics which are, quite often, unreliable). Migration patterns may necessitate radical changes in project design and scheduling of project works. For example, in both West Africa and Nepal there is a high percentage of seasonal male migration, for work and for social reasons. In these circumstances the project has to take into account these patterns and might have to re-schedule project works, or consider mechanization. The project would also have to determine whether those left behind (say, women) are decision makers who can continue to implement the project in the absence of the others. At times migration may alter existing patterns. For instance, the out-migration of males in Swaziland to work in the neighboring mines in South Africa has allowed the matrilineal lineages to consolidate their control over land and to oust males from their traditional rights of cultivation in lineage plots (Gloria Scott personal communication).

The second question is whether there are any traditional forms of labor exchange and cooperation among the project population which can be used to cultivate new crops that the project proposes to introduce; and conversely, whether these traditional forms will hinder attainment of proposed project goals. Labor exchange and cooperation are quite common traditionally, e.g. the *gayap* (Trinidad), *donkpe* (Dahomey), *gandu* (Nigeria), *gotong royong* (Indonesia), and *nogar/parma* (Nepal). Few projects consider the impact of these traditional forms of labor exchange on project goals, or attempt to assess whether traditional forms of land use can be employed in attaining project goals.

There is increasing evidence which permits, with appropriate cautions, two generalizations with regard to the use of traditional forms of labor exchange for the cultivation of new crops: first, the bureaucratization of traditional forms (or their formalization into "rules") results in a loss of meaning to the participants; the traditional form loses its effectiveness, and gradually ceases to operate. Second, when new crops have been introduced, traditional rules with regard to the division of labor and who cultivates which crop do not appear to apply (Sossah 1980). There also appears to be a similar relaxation of tenurial rules when a group migrates and new rules are formulated (Hill 1963). For the project designer, it is important to assess the extent to which the introductionm of a crop will affect traditional rights--for instance, oil palm in Nigeria was planted on "women's lands" (that is, lands traditionally allocated to women), thereby effectively depriving them of their traditional land allocation and the income derived from these lands.

Security of tenure

What are the rights of the persons using the land?

In this section, two questions will be considered: first, in what capacity does the person use the land? And second, what is the duration of possession?

The capacity in which a persom uses land is extremely important since it influences the duration of land use as well as eligibility for credit. Access to lands is determined by membership (real or fictional) in the group. But once access has been obtained, then the duration of land use relates to the status of the person--is it a son or a wife, is the cultivator a sharecropper or a tenant, is shifting cultivation practiced? In effect, what must be asked is what security of tenure does the person in actual possession of the land have?

Transfers of land do occur in Ghana, among the Ibo, and in the Gambia. Thus, it is not quite accurate to lump together an entire region (West Africa, for example) and claim that because land is "communally" owned the person in possession cannot mortgage, rent, lease, pledge, or enter into an agreement for conditional sale of the land.

Different considerations arise where there are tenants on the land. Some tenants may be protected in possession by law (as in India), or by tradition (as in Haiti and Brazil); others may not. But, if there are tenants, project design must take them into account lest the benefits go only to absentee owners (Sorbo 1980). Further, as in the case of sharecroppers (a recognized feature in Latin America, Tunisia, the Sudan, Madagascar, Turkey and India, to mention just a few countries), consideration must be given to whether occupants of the land will have any incentive to double-crop, or use project inputs. In the case of irrigation projects in Madagascar, for instance, it was noted that sharecropping, though illegal, still continued, and that there was no incentive to these cultivators to double-crop in the projects.

What happens when the person in occupation dies? Does the land revert to the group which, acting through their leaders, can reallocate the land to some other person not necessarily related to the former occupant? The right to land is usually heritable, but the manner of inheritance differs. This means that an assessment must be made of what is likely to happen if the project is implemented. For example, what may appear to be sufficient land for a family unit at the time of project implementation may be insufficient, or may affect project economic viability, when inheritance laws are considered. For example, in India, in countries where Islamic law prevails, and in Papua New Guinea, each succeeding generation inheriting a piece of land can demand partition and promote further land fragmentation. In these circumstances, project design should be able to take into account this traditional division of lands and assess whether project targets can be attained, or whether new rules should be framed modifying traditional inheritance rules--something that has been attempted, not very successfully, in settlement projects.

The second question, the duration of use, is linked with the capacity of the person using the land. Duration is quite often divorced from ownership of land. Thus, for example in Lesotho, although land is said to be "communally" owned, surveys have shown that farmers held allotted lands for nearly their entire farming careers (Eckert 1980). So, too, among the Lozi (Zambia) where, theoretically, the cultivator had only the right of usufruct (Gluckman 1965). A similar situation exists in Nigeria and Ghana. But where the cultivator is a sub-allottee (a wife or a son) or a sharecropper, the duration of tenure is much more tenuous and tends to exist only as long as the person remains a wife or son, or until the superior holder evicts the sharecropper.

Beneficiaries of the land outputs

How is the produce of the land distributed?

It is a mistake to assume, as is often done, that the unit that produces the crops (an extended family, for example) also consumes the crops. Thus, for example, among the Sere of Senegal, harvest shares go by custom to heads of compounds, wives, religious authorities, and heads of the matrilineage. In the Indian joint family, a similar division of harvest shares takes place among the male members of the family (and their offspring), despite the fact that only one brother may be cultivating the land. Again, where further fragmentation would lead to uneconomic holdings, the practice (in India, Turkey, Sri Lanka, and Brazil, for instance) is for members of a family to cultivate the land in rotation and for all others to receive a smaller share than the actual cultivator. In these circumstances production decisions are divorced from consumption decisions. For the project designer, then, the questions that must be answered are: Would the actual cultivator agree to use proposed project inputs to increase production? What design of project organization and communications components would be necessary to reach and influence both sets of decision makers?

Other operational questions

There are two other considerations, related to traditional patterns or aspects, of land use, that should be evaluated before project design is finalized. They are: Is the population in the proposed project area multi-ethnic? and What is the carrying capacity of the land?

Is the target population homogeneous or multi-ethnic?

Project design often assumes a homogeneous population and that all persons residing within an administrative unit, say, a village, form a "community." There is a polite tendency to shy away from references to known facts, such as the multi-ethnicity of the population or the fact that there is caste stratification. And yet, the existence of multi-ethnicity or caste stratification could affect the attainment of project targets because of difficulties in obtaining common participation or competing claims for project benefits. Particularly in settlement projects, it is necessary for the project designer to enquire whether the immigrant groups would be accepted by the settled population. Where there is caste stratification, there is also a division of labor based on caste. This would affect labor scheduling for project works, as well as assumptions regarding the distribution of benefits among the target population.

Carrying capacity of the land

It is necessary to estimate land needs, particularly in projects relating to transhumants and shifting cultivators where observations made during one season of a year, or even during an entire year, can be wildly inaccurate. Land needs, again, must be estimated against likely population and climatic trends, the level of technology currently used, the technology which the project proposes to introduce, and the time within which the change might take place. Also, whether project incentives would lead, perhaps unintentionally, to overcropping and mining of soil fertility or accumulation of larger herds and consequent overgrazing.

To sum up, an answer to the question "who holds what interest in land?" must involve an examination, on the ground, of the nature of possession: who is in possession of the land, what are the rights of the occupant (security of tenure, ability to deal with the land), how did the occupant obtain possession of the land (allotment--whether by an individual or group), how is the land used (type of crop, division of labor, competing groups/uses), and how is the produce of the land appropriated? In all cases, there is an identifiable group or individual that makes decisions on allotment, use, and appropriation. Decisions with regard to allotment and use are usually arrived at in accordance with principles that reflect both the philosophy of the group and the internal distribution of power. If the proposed goals of the project are to be attained, answers to these questions and an understanding of the traditional system are essential. Project goals should be in consonance with the traditional system and, where they are not, it is only an understanding of the system that would provide the designer with clues as to where changes will be necessary.

3. Land tenure issues in the social/community forestry project context

In these projects, if the trees are to be located on community land perhaps the most fundamental question is the availability of land. What does this mean? Let us suppose that the project planner does his "homework." He has studied all the statistics available. He discovers that inequalities in land holdings are a fact of life; that, let us say, 25 percent of the villagers in the area selected are landless, 40 percent have holdings of less than 2 hectares, 30 percent have holdings of between 2 and 5 hectares and the rest have holdings above 5 hectares. He wants to reach the lowest 65 percent. There are several further "facts" he must find out. Do the landless have enough land to grow even a few trees, are they too busy earning a living (through labor wages) to have time to plant and care for trees, if seedlings are given to them will this cause resentment among the 35 percent that will not get seedlings under the program? For the 40 percent (with holdings

<2ha) that are also in the target group there are further questions: What is the nature of the title to the land they farm? Are they tenants, sharecroppers? Is the area one where shifting cultivation is practiced? If so, is there any likelihood that the farmer will move away from the farm before the trees reach maturity? If farming plots are reallocated every few years, is there any chance that the farmer who grows trees will get the plot back?

Other questions must be considered if the goal is to develop community forestry on common lands. Is there enough land for this purpose? Often, the availability of sufficient land is only theoretical--it is to be found on the land records, or the government says that the land belongs to the State (statutory ownership). As anyone who has been in the field knows, there is an enormous difference between what is on the books and what happens on the ground. The common land may have been encroached upon by the village elite or the landless. The land may be used by different groups with unequal power: the village herders, the transhumant grazier. For the former group the land may have little value because they send their herds with a "master herder" to other lands for grazing or they may have sufficient fodder (from trees or agricultural residue) to feed their herds. The altered use of the common land is of little consequence to them. Even between members of the village there may be different effects if common lands are put to altered use and the lands closed, even temporarily until the plantation is established: some may need the common land, others may be little affected if the land is put to trees. How can those who will be most affected be convinced to join in the common venture, or at least not destroy the plantation? "Oh," the extension forester says, "just wait for two years and you will have grass, at year three you will start getting twigs..." And the villager replies, "I cannot wait for a year. Who is going to feed my herds? Must they die? I have no labor to take them to other grazing grounds, nor can I afford to pay someone to do so." Does the forester then say, "I will give you wage labor and with the cash you can purchase fodder"? Suppose the answer is "there are no fodder markets"? What alternative arrangements can be made for the period before the plantation is established?

We have so far assumed that there is sufficient common land and that the problem is the different uses of land by groups within the community. What, however, is meant by "sufficient" land? The meaning depends on the goal of community forestry in a particular locality. If the community plantation is only meant to serve as a "demonstration" of what trees can be grown, then small (say one or two hectare) plantations would suffice for the purpose. Most farmers, know the local trees; therefore demonstrations should emphasize exotic species. If the goal is not merely demonstration then two conditions should be satisfied: the area selected should be large enough to provide a useful amount of fodder, fuelwood and timber; second, the species mix should include exotic and known species, with the balance tilting in favour of the latter.

Whether or not the community plantation is intended as a demonstration plot, three other matters have to be settled. First, the labor sources for pitting, planting, maintenance and harvesting. Second, the distribution of produce. Third, whether any charges should be levied for produce and whether any payment should be made for labor involved in plantation operations. Many projects seek to resolve the problem of labor by avoiding the question. The Forest Department is left in charge of operations: the Department employs labor for plantation establishment and protection; the Department harvests the plantation and distributes the income from harvest or produce after deducting costs. The village then allocates the income for its own projects. This has been the case in many projects in India. An important question is whether this approach transfers any forestry skills to the local population, and whether it even serves as a demonstration vehicle. Even more importantly, the local population may get the impression that the project is a "government plantation" in which the stakes of the village are low. The villager does not identify with the project. Assuming, however, that local labor is to be used, should wages be paid and who should fund the payments?

If you listen carefully, you can often hear the following statements in the field. Local leader to Forest Department official: "Oh, yes, we have sufficient labor for the work. We have done many voluntary projects before and this project will bring such benefit to the village." The official goes away, happy with hearing what he wanted to hear. The laborer (whose services have been commandeered at short notice): "Another voluntary project. Why is it that only the poor sections of this village have to 'offer' voluntary labor? We never get paid. I now have to work for three days, just because I owe money to the village chief and need to work in his fields. No payment. After I finish working, I will be forgotten." Quite often "voluntary" labor is not voluntary. It is the poorest villagers who are called upon to work, which they cannot afford to do. When the project is complete, the village leaders pose for photographs with the officials, everyone but the laborer is happy.

Should wages then be paid? If so, from which funds. These questions are not easy to answer. If the funds used are government funds, the villagers might get the impression that the plantation would in reality be owned by the government and that all the statements about the plantation belonging to the community are not true. Then, again, the community may have insufficient funds to pay wages; or they may need those funds for other purposes. Before any final decision is made about payment one question must be examined carefully. Does the community have any examples of successful community projects? If so, why did they succeed? Who received the benefits? How were decisions arrived at? Who participated in decision-making, work, and wage payments? What was the nature of the project? In other words, did the community rate the project as one of its highest priorities? If there are no successful examples of community projects, efforts must be made to find out why. It is only then that a final decision can be reached to whether or not payment should be made.

A final decision which must also be made very early, even before a community project is started, is the distribution of benefits. Many who are outsider-advisors try to prescribe their own notions of equity in such distributions. The advisor comes in, talks to officials who immediately accept the advior's notions, and goes away convinced that a change has been accepted. Is this realistic? Does the advisor really know what happens after he leaves? Do mere conversations change the power structure in the community? The officials merely chuckle. They talk among themselves, "By the time the period for distribution of benefits comes, the project will be over. No one will come and find out what has really happened." That is why it is so important for the forest extension worker to talk to different sections of the population, to find out from each what happens, to get all sections involved in arriving at a principle of benefit distribution. If produce cannot be distributed, then on what will the income be spent? Will the object of expenditure benefit all sections of the community, or only a few sections? The general rule in community projects is that those participating must be convinced that they will get something from the project. If there is no conviction, there cannot be a community project. To ask "what is in this for me?" is not uncommon. The question must be answered.

4. Land tenure issues in the agroforestry project context

In agroforestry projects, trees are planted in various combinations with other crops, often on individual or household holdings. The close relationship between land tenure and trees in these circumstances is well documented in the literature, in a surprising variety of circumstances. In densely populated areas of Latin America, maldistribution of land and resultant pressure on land have forced clearing of tree cover from hillsides with disastrous environmental consequences (Eckholm 1967). In pioneer zones, peasants clear land in an attempt to establish title but then lose it to ranching interests attracted by the cleared land (Denevan 1982, Jones 1982). Where peasants lack secure tenure they may be discouraged from planting trees (Sellers 1977). In Africa, traditional land tenure systems that confer important land management rights on a lineage or community may create disincentives for tree planting. In some communities, trees may revert to the clan on the death of the person who has planted them, rather than to his or her family (Rald and Rald 1975). Where tenants or others are using clan or community land by permission, they may be barred from planting trees as an act inconsistent with their precarious tenure (Shambi 1972, Tanner 1960, James 1971). Even tree planting by those with allocations in their own right may be resisted as interfering with the rights of reallocation of chiefs or other traditional authorities, and in some societies prior permission is necessary to plant trees (Brain 1980, Duncan 1960). Indeed, the African literature is replete with examples in which, having been allowed to plant trees, holders with temporary or fragile titles have enhanced their tenure, usually at the expense of the group. Not uncommonly, concerned observers conclude that existing tenure systems pose an unacceptable degree of ambiguity and that private rights of landowners in trees must be strengthened (Thomson 1982).

Agroforestry initiatives raise many of the same tenure issues as those associated with the shift to cash crops in response to market stimuli. But they raise special issues as well. Trees are late maturing crops and investment in them may require greater security of tenure than shifts to new annual crops. Also, trees commonly increase the value of the land upon which they have been planted, and thus raise land tenure issues resembling those raised by other permanent or semi-permanent improvements in the holding. These similarities and distinctions are developed below.

Shifts to new crops, with new inputs and new farming practices, raise tenure issues of varying seriousness, depending upon the circumstances. Such shifts may be retarded by tenure arrangements. The introduction of new crops, intended primarily for marketing rather than home consumption, raise a land distribution issue. A small size of holding due to maldistribution will often make it difficult to reduce production of subsistence crops to produce a new crop for the market In this connection agroforestry may offer distinct advantages over conventional cash crops, since it is often possible to integrate cash crop trees with subsistence crops without displacing them. In addition, it is important to consider whether the farmer who will have to bear the cost of the investment will receive the full benefit of his investment. If he will have to share it with other rightholders in the land, his incentive to invest may be reduced to the point where he will not make the investment. For example, in many traditional African tenure systems when lineage members leave the village to work in town they may, in principle, retain their rights in lineage land even though it is worked by someone else in their absence. The issue here is not so much insecurity of tenure as ill-defined obligations to the absent party (see also Bruce 1983).

Other tenure issues are raised by the changes in land use which new crops may require. These may include a longer or a different growing season, or even double-cropping. Serial land use by primary and secondary right-holders may cease to be possible. If project planners incorrectly assume that a primary user has exclusive rights to his parcel, it may lead to serious problems. A common instance involves innovations inconsistent with continued use of farmland as community pasture after the growing season. When the community presses its claim, it can, for instance, prevent adoption of a new tree crop which is not compatible with the use of the fields as pasture. The issue may generate considerable heat; community leaders may feel that the principle of community control of land is at stake. Such conflict may arise not only between individuals within a community but between communities and ethnic groups. Serial use of land by two communities may be incompatible with the innovation concerned. In this form the issue can have serious political implications (see, for example, Deng 1984). The serial uses involved are usually farming and grazing, and the common reaction is to ask how to provide more exclusive and secure tenure to the farmer. The problem will prove intractable, however, unless equal attention is given to finding alternative feed resources for livestock owners.

The land tenure issues raised by these project components can be greatly magnified if the new inputs make possible for the first time a transition from shifting to permanent cultivation (see Raintree, this volume). In West Africa and many other parts of the world this has been intimately associated with the introduction of commercial tree crops (Hill 1963, Berry 1975). Elimination of serial land use is one consequence, but there are other issues as well, which transcend problems of particular individuals and involve broad systems maintenance and reform issues. How will the community cope with the new stability of landholding and provide secure tenure? Once agriculture begins to sedentarize, relatively intense competition for the better pieces of land can arise quite rapidly. Do there exist adequate institutional resources to cope with dispute settlement within the indigenous system, with appeals into the modern court system? Inheritance of land under shfting cultivation may have been an undeveloped area of customary law. How can one ensure elaboration of a law of inheritance of land which avoids uncertainty? Other new legal mechanisms may also be needed as land holdings stabilize, such as land borrowing or leasing to adjust temporary disparities between household landholding and labor supply. A project which aspires to promote or operate within the context of this very fundamental change in agriculture must consider how these needs can best be met.

New crops and other new inputs generally involve some investment, but that investment--and the land tenure issues raised--are usually substantially different from those involved in major, long-term investments in the holding. Trees, however, are a long-term investment and some of the tenure issues raised may be more like those raised by major investments in the holding than by other changes in crops. First, the investment may be large, and this makes any risk more serious. Second, a large investment will mean that costs can be recouped only over a number of years. The period of exposure to risks posed by lack of security of tenure is increased, and risk is more difficult to estimate in the longer term. Third, vulnerability to risk is increased by the merging of the investment with the holding; it cannot as a practical matter be satisfactorily detached from the holding if the holding is lost, and will often be legally inseparable from the holding. Fourth, risk is increased because of the anticipated or actual increase in the value of the holding, inviting land-grabbing and competing claims by the powerful.

Such insecurity can thwart project objectives in three ways. The farmer may decline to make the investment in his holding, considering the risk too great. Alternatively, he may make the investment and then find himself displaced by more powerful claimants, attracted by the value of his improvements. Or he may lose his land before he is even aware that he has a choice to make (displaced in anticipation of the project).

Longer term improvements in the holding can also, in certain circumstances, antagonize traditional authorities. This is because the improvements seem to assert a permanency of right in the parcel, in derogatiion of the community's right to allocate land, and also because such investments portend the growth of a competing class of "big men." This antagonism may find expression either directly through reallocation of the land or, more commonly, indirectly through harassment (such as accusations of witchcraft), aimed at forcing the farmer to move from the community and abandon his land.[7]

Let us summarize tenure issues raised by tree crops, as new and long-term investments in a holding. An investment in trees will be difficult if maldistribution of holdings has left many farmers with only enough land to meet subsistence needs. An investment in trees will also be discouraged if the farmer bearing the cost of the investment will not receive the full benefit of his investment, but will need to share the benefit with other rightholders in the land such as a landlord or co-owners. Trees represent a major change of land use, one which may often be prohibitive of other, formerly serial uses. Tree-planting thus has potential for creating conflicts among members of the same community or between communities. Insecurity of tenure will under most circumstances discourage the planting of trees. Because trees may make virtually valueless land quite valuable, tree-planting may attract competing claims to the land, claims that may have long lain dormant or claims simply fabricated by the powerful. Success may breed envy and unreasonable demands from traditional leaders. A farmer who grows trees needs exclusivity of tenure and security of tenure. And, because trees are long-term crops, their treatment under laws of inheritance must be clear and positive.

An exception exists to the proposition that insecurity of tenure discourages tree planting. It seems at first glance a rather specialized one, but it is apparently not uncommon. In certain circumstances, trees may be planted to create or preserve tenure, to peg down a holding. Commercialization of a tree crop may provide the incentive, and the establishment of plantations can be a critical step in the more general transition from shifting to permanent cultivation, with trees defining permanent holdings. This process has been well documented in Mexico and in West Africa, where it occured in spite of the fact that many of the tree-planters were initially "strangers" holding the land by permission (Foster 1966, Millon 1955, Hill 1963, Berry 1975). There is, of course, a question of how many trees one needs to plant to establish land rights. A simple border of trees around the property may not be sufficient to establish a claim to the land (James 1967); a certain density of trees may be required before the claim is accorded legitimacy (James and Fimbo 1973).

Trees take advantage of the almost invariable customary rule that land under crops should not be reallocated, and replanting makes the right virtually perpetual. Tree planting may initially be discouraged by insecurity of tenure but, at least in certain circumstances (which are not well understood), once there emerges a sense that rights in trees will be protected, tree planting may actually be motivated as an antidote to insecurity of tenure. Some odd results occur in project contexts. In Bong County in Liberia, for example, farmers seized the opportunity to grow project-supplied cocoa and coffee seedlings partly because the trees would establish their use of the land and prevent the government from selling it to speculators from Monrovia, and because the farmers assumed the project donors would exert pressure to keep the trees in their hands (LTC 1982). In Ivory Coast, farmers rejected government efforts to get them to intensify cocoa production because land-extensive cocoa production allowed them to claim larger areas of land (Tiffen forthcoming).

Much of what has been said so far assumes that trees belong to the person upon whose land they grow, and at this point it is important to recognize that that assumption will not always be valid. In Western law, trees are "fixtures," items "permanently affixed" to land and assumed by law to belong to the owner of the land (unless, of course, there is a contract specifying the contrary). But it is not unusual for indigenous systems to recognize "tree tenure" as being at least partly distinct from land tenure. A tree may be owned by someone other than the owner of the land, or may be co-owned by the land-owner and, for instance, the person who provided the seedling and planted it along with the person whose pump has watered the tree. This is the case with date palms in the northern Sudan, where the Islamic rules of inheritance can often result in co-ownership of a tree by several dozen persons, only a few of whom have any claim to the land on which it stands. See Fortmann (this volume) for a review of tree tenure in customary law. The intention here is simply to stress the importance of this phenomenon and to pose the question of whether, in a situation of insecure land tenure, tree tenure can offer an alternative basis for economic security.

5. Dealing with land tenure issues in the project context

Options for response

Let us assume that we have a project idea and that it has become clear to us, fairly early in the planning process, that the project model we have in mind is not going to mesh smoothly with the tenure pattern of the intended project beneficiaries. The problem that we foresee is one which appears likely to seriously discourage the intended beneficiaries from taking advantage of the opportunities posed by the project. What is the range of options available?

Avoid the problem by relocating the project

Potential tenure problems are oftem more acute in some localities than others. A recent AID project paper from an African country notes the seriousness of tenure problems there and indicates that the mission has been able to proceed with its agricultural programs only by "judicious selection of project sites" (USAID/Mauritania 1981). As that paper acknowledges, however, this is often not the best approach. It may be justifiable, nonetheless, if the tenure problem concerned is quite special to the project area first considered. For instance, a project is shifted from District X to District Y because it has been realized that the provincial governor's family owns most of District X and claims more. If this is an unusual circumstance, it is not unreasonable to look for a new project site. But consider another example: intense pressure on land has produced a high level of land litigation and consequently great tenure insecurity among farmers in most parts of the country. A project site is therefore selected in which size of holding is three times the national average, and where the trends noted elsewhere do not yet threaten farmer investment in their holdings. In this case the project is seeking out an atypical situation and replicability is lost.

Reframe the project so that the problem does not arise

This involves finding other ways to offer the opportunities of the project, ways which permit beneficiaries to seize those opportunities consistent with the existing tenure pattern. This will be the most attractive option in many cases. Land tenure is not easily changed in the short run, and there is no point in undertaking the task if a little imagination makes it unnecessary. For example, an agricultural credit project is based on a model utilizing land security for loans. Noting tenure obstacles to securing loans with land, and having reviewed the credit needs of beneficiaries and found they are relatively modest and short-term, the project planners substitute crop loans and cattle mortgages. See Bruce (1981) for a discussion of alternatives to real property security for loans in Botswana. It is not always easy, however, to find a way to reframe a project so that it is consistent with existing tenure patterns. In some cases tenure change may be required.

Modify the land tenure pattern so that the problem does not arise

If the tenure situation of concern cannot or should not be avoided by shifting the project site, and the project idea cannot be adjusted to the tenure situation without seriously undermining the project rationale, then tenure modification deserves consideration. There are two possible levels of response.

First, it may seem appropriate to attempt to deal with both the tenure issues raised by this project and other problematic aspects of the tenure system in a comprehensive and relatively conclusive manner. Land tenure reform becomes a major part of the project idea.[8]

Alternatively it may seem inappropriate or infeasible--for reasons ranging from political obstacles to limited funding--to deal with the tenure issues in a comprehensive and conclusive way. In that case one does just enough to protect the project and its objectives. This approach may prove shortsighted, but in some circumstances it will be the only option. Often one combines this with the previous option, altering the project idea a little and altering the tenure situation a little, making the minimal adjustments needed to achieve a workable accommodation.

Drop the project idea

This is a course of action to be considered only after the above possibilities have been explored and found unworkable. But it is important to affirm that it is an appropriate and sometimes necessary conclusion to the examination of a mismatch between a project idea and a tenure situation.

Tools of tenure reform

Of the options mentioned above, tenure modification is of particular interest to us here. How is it achieved in the project context? Several instruments of tenure modification can be distinguished: a) national legislation, b) community legislation, c) "the law of the project," d) contract, and e) project economic leverage.

Several of these are categories of law, and it is useful before discussing the particular categories to briefly consider law as an instrument of social change. It is an important element in most tenure reforms, but it is neither a necessary element in every tenure reform nor is it a sufficient condition of tenure reform. Its usefulness is enhanced if its limitations are appreciated.

Under what circumstances can we expect people to obey a legislated command that landholdings "shall" be managed in a particular way?

a) When most people are already behaving in that way, in violation of existing law or simply in a legal vacuum, or wish to behave in that way; law reform acts as a ratification of practice or a release to act as desired;

b) Where some people are directly interested in the new rules and thus motivated to enforce their new rights, and where they are given effective civil remedies in an accessible and reliable dispute settlement system;

c) Where the state itself actively monitors and sanctions the breaking of the law, as when criminal sanctions are imposed; and

d) Where the state's administrative machinery directly intervenes to change facts (for instance, replanning parcels and layout, consolidating holdings, creating local administrative machinery), which so alters the position of most people that they have little choice but to comply with the new set of rules which accompany the new facts.

It needs to be emphasised that it is not law alone that changes behavior, but law which authorizes, organizes and releases other forces which, acting in conjunction with law, change behavior. While they are discussed separately below, they are usually effective only in combination.[9]

National legislation

Most agricultural development projects are not implemented on a national level. They are more commonly regional or area development projects, and this may be due to one of several factors: 1) the project is only appropriate in certain geographically limited resource situations; 2) the project is experimental, seen as a pilot effort; 3) the project is expensive, and therefore implementable only in limited areas in the near future; and 4) services needed to complement the project, for instance, marketing infrastructure, are unavailable outside the project area.

The non-national nature of most projects becomes problematic in the tenure reform context because frequently a critical component of tenure reform is law reform. In many cases law reform is a necessary first step in the tenure reform process, altering rights and duties between persons over resources before setting out a plan of implementation to realize the new rights and duties and creating administrative machinery and institutions to pursue that plan.

Legislation is almost exclusively a national prerogative in Africa. Land tenure, on the other hand, is a profoundly cultural phenomenon and varies with ethnicity. In multi-ethnic states, which include most states in Africa, different tribal areas have different tenure systems and tenure problems. In addition, even if tenure systems are relatively uniform, conditions creating a need for reform may exist only in limited areas, as when a project creates that need.

It is technically feasible to frame a tenure reform in a way which recognizes and is sensitive to the particularity of tenure systems. Regionally applicable legislation would meet with no fundamental jurisprudential or constitutional objection in most African countries. But such a legislative proposal may run into political difficulty. That difficulty has two sources. First, a program which adjusts itself to or seeks to address directly the particularities of a tribal land tenure system may be seen as pandering to and reinforcing tribalism, or at least perpetuating divisive particularity. This is a concern at the level of the national political elite. But there is a second reaction which may come from the people directly affected by the project and their political representatives. Tenure reforms are generally greeted with some suspicion by those directly affected, often even by the intended beneficiaries. Those suspicions multiply when local people realize that they are being treated *differently* from others in the nation. "Why us?" is the cry, and paranoid scenarios are unfolded. In pre-revolutionary Ethiopia, a set of land reform bills were stalled for years in Parliament partly because of such reactions.[10]

Legislation can be framed which seeks to meet such objections. For example, the empowering legislation can be framed in relatively general terms nationally, with considerable authority given to government to prescribe different regulatory rules for different localities. This will be a workable approach in some situations but it is subject to one disadvantage: at least in parliamentary systems, rights based on regulations (which can be changed by government at will) are not so secure as rights created by the law itself.

Alternatively, the empowering legislation may enact a national program but provide for its implementation only in particular areas, gazetted from time to time by government as standing in need of the program. This last is obviously most appropriate when emerging tenure issues are the same nationally and the problem is one of geographically limited immediate need. Of course, it is possible to combine the three elements: a nationally uniform frame, diversity in detail, and gradual implementation. Whether these approaches to legislative drafting for meeting local needs will be acceptable will differ from situation to situation, but they are options which deserve serious consideration.

Community legislation

In most African countries government is highly centralized, with an institutional void between the national center and local communities. There are not many opportunities for legislation at, say, provincial level. To find an alternative to national legislation it is often necessary to go down to the community level. "Community" here is used loosely--the intent is to refer to the level of the indigenous society at which key decisions about changes in customary law are made. That level will vary from place to place.

Such decisions *are* made, and there is such a thing as community legislation. "Customary" tenure is not simply rules deeply internalized, obeyed by ancestors from "time out of mind;" nor does it change only according to the generally invoked model of incremental change. In that model, deviance increases until it becomes pervasive and, in time, a new custom is recognized. The process is depicted as almost unconscious. Certainly much change in customary rules takes place in this way, but it is not the only way. When confronted by drastically new needs, traditional communities (tribes, clans, villages) sometimes purposefully decide on rule changes and, in effect, legislate changes in their customary land law.[11] One strategy that deserves serious consideration is the encouragement of such community legislation. Projects are able to promote such decisions either through simple encouragement or more directly by preferential treatment of villages which have taken the appropriate steps.

This approach can be labor-intensive and time-consuming. Often the response of local communities is difficult to predict. In addition, enforcement mechanisms (the national courts system, other institutions) must be considered; community legislation needs backstopping and local institutions may not be able to provide it. But this approach has three major advantages. First, it involves local people in making decisions that affect them, and thus develops human resources. Second, this participatory nature is good insurance against mistakes, especially project management's failure to anticipate unfortunate side effects of change. Third, it permits a degree of local experimentation, with different communities adopting slightly different solutions to the problem.[12]

"The land law of the project"

Sometimes there are tenure decisions that a project itself is legally competent to make. This is most clearly the case when the project area or a critical part of the project area has been State land or has been acquired by the State specifically for the project. In those circumstances, a project can virtually create a "law of the project." This is done through regulations and standard contracts, often incorporating special development conditions, requirements of compliance with a common plan of management, and even special inheritance rules. Project-developed land tenure in settlement projects, irrigation projects and pasture management projects relies heavily on these mechanisms. Legally, the situation is one in which the project has almost total discretion. As the State's agent, it is the owner of the land and it can impose a tenure system as a condition of access to that resource (see Bloch 1985, Tiffen 1984).

This approach works best where the land concerned has been previously unoccupied, or its character has been dramatically altered by the project. In the former case there are no established expectations to disturb. In the latter, expectations may more readily be adjusted because of dramatic physical changes in the resource base. When working with settled communities, however, the authority of the State as landowner should be exercised with some caution. Commonly, those farming will not have been told--or be reluctant to accept--that they have no rights in the land they farm. A heavy-handed approach by the project can create resentment and mistrust, and may sometimes be fundamentally unfair.

Contract

The notion of a "land law of the project" includes contracts which are imposed by the project as a condition of access to land. Contracts are utilized as a tool of tenure change in other circumstances, where government is not the landowner. For example, an agroforestry project encourages landlords and tenants to enter into leases with fixed rents and extended terms, as a condition of access to project credit. Or a group of villages contract with one another concerning the gathering of fuelwood in a common forest. In these cases, contract is used as a means of creating the desired framework of interaction between individuals.

Alternatively, contracts can be utilized by projects as a tool for regulating tenure relations between groups. For instance, a boundary disagreement between groups may be best dealt with through mediation, and the result expressed in the form of a contract between the groups. There are also contracts between a project on one hand and groups and individuals on the other. A social forestry project may seek to avoid tenure insecurity with respect to its own community woodlots by obtaining a long-term lease from the community. Or a project may join as a party to one of the earlier mentioned types of contract, promising assistance to the parties as an incentive for them to compromise and settle their dispute.

Contract is valuable to project planners as a mechanism for restricting behavior or requiring certain actions beyond what is required by law, and creating obligations that will be enforced as law by the courts. It is in theory a form of legal obligation assumed only voluntarily by the parties, but in fact project planners can sometimes arrange offers that can't be refused.

Project economic leverage

Projects can affect behavior with economic leverage exerted through preferences, subsidies, and a wide range of other actions. Most often, these will be used in conjunction with legislation, which authorizes and organizes the economic incentive or disincentive, or with contract, which can be used to create the legal obligation which is the *quid pro quo* for the incentive. Many of the examples given in this section have involved some combination of legislation, contract and project economic leverage. But project economic leverage can be used in ways which do not involve legislation or contract, just simple material support to particular interest groups in the community. Village X, for example, seeks to establish exclusive use and better control over an area of forest being damaged by herds; the project builds a fence for them, and supports the chief in his efforts to impound trespassing animals. Or in an agroforestry project, extension and marketing services are provided only to villages in which councils of elders decide to allow fencing of holdings.

6. Research goals

In conclusion, we want to suggest several issues which appear to us to merit further research:

Earlier in this paper, an attempt was made to sketch out what we know about insecurity of tenure and the planting of trees. It is an arguable hypothesis that under at least some African indigenous tenure systems, defining the problem as one of insecurity (except in the case of tenants and strangers) may be misleading. It may be that there is simply an initial and relatively reflexive and weak resistence to be overcome, a resistence based in traditional authorities' reluctance to see a situation arise in which their reallocative function will atrophy. In certain circumstances, it has suggested, insecurity may encourage tree planting (as an antidote to the insecurity). We have as fragmentary evidence in support of these hypotheses a few case studies strung out along what appears to be quite a broad spectrum of experience. We do not have enough data to specify with any confidence the circumstances under which indigenous tenure arrangements encourage or discourage tree planting, and future research should aim to elaborate through comparative study a scheme which permits such specification. This could best be done not in isolation but as part of a broader inquiry into what motivates farmers to plant or not to plant trees, a study of incentives and disincentives.

Tree tenure emerged during the planning of this workshop as a major concern, and Professor Fortmann's paper clearly establishes the importance of the phenomenon. What is less clear from the data available at this point is the historical, perhaps evolutionary, relationship between land tenure and tree tenure. For instance, might it be that tree tenure originates for scarce economic trees in circumstances in which land scarcity has not yet led to long-lasting household or individual property rights in land, and is gradually replaced by a "fixture" approach once such land rights develop, perhaps lingering on with respect to planted trees in common areas? This is hardly a hypothesis which does justice to the complexity of the data, but it will give some sense of the kinds of relationships future research should seek. This inquiry is directly connected with a policy issue of considerable practical significance: is tree tenure, disassociated from land tenure, a viable option for providing the security and certainty of expectation required as incentives for tree planting and husbandry?

Another area not adequately researched to date concerns the impact of laws of inheritance on tree farming on individual or household-held land. Certainly the impact of such laws on the holding of trees can be bizarre--witness the extended co-ownership of date palms under Sharia in the Western Sudan--and can have an impact on tree-planting, as in the case of "family lands" in the Caribbean. There is a need for more detailed studies in this area, to build up the information base. Two possible foci are the Sharia rules of inheritance and those in matrilineal societies, but others of equal importance may be suggested by the participants in the workshop.

Notes

1. A tree crop settlement scheme in Western Nigeria failed because settlers were not given adequate security of tenure (Ogundowara and Heady 1973).

2. AID's Bong County Agricultural Development Project, which has emphasized coffee and cocoa planting, was begun in a situation made for beneficiary displacement. Traditional holdings had little legal security on land which was legally State Land and subject to sale by the State to elites at nominal prices (Harbeson et al. 1984). Also in Liberia, a rural roads project found that purchases of land by elites and urban dwellers along the planned routes had created a wide barrier between the roads and the small subsistence cultivators whose marketing and other needs the roads were to meet.

3. The phrase "interest in land" is used in a wider sense here than in law where, for instance, a licencee has no interest in land. In this paper the main concern is occupation, and the terms of occupation, of land.

4. In shifting cultivation field boundaries are not maintained and the fallows are generally "stages in a return to a well-established woodland or forest, within which the signs of earlier cultivation may be difficult to see" (Morgan 1969).

5. Under an agricultural development project in Trinidad and Tabango farmers "lost interest" in tending citrus plants because they would not bear fruit for five years. For more on social factors in forestry projects see Noronha 1981.

6. Carrying capacity "is the number of people and the level of their activities which a region can sustain in perpetuity at an acceptable quality of life and without land deterioration. . . Carrying capacity recognizes that all habitats have a finite assimilative capacity. Theoretically when this upper limit has been exceeded, a gradual environmental decline will set in, ultimately producing adjustments to the population itself" (Bernard and Thom 1981).

7. The Land Tenure Center is beginning a study in Swaziland in the fall of 1985 which will hopefully explore the mechanics of this process.

8. AID's Ada Agricultural Development Project in Ethiopia in the early 1970s became directly involved in testing models for reformed landlord-tenant relationships, in part to avoid widespread eviction of tenants due to mechanization sparked by the project.

9. There is considerable literature on law as a a tool for inducing economic and social change in developing countries. For example, see Seidman (1968).

10. In the early 1970s, the government was proposing to Parliament a bill to regulate landlord-tenant relationships (most problematic in southern Ethiopia, and bitterly opposed by landlord interests there) and a bill to register holdings (most problematic in northern Ethiopia, and opposed there as fatal to the area's communnal tenure systems). Parliamentarians from the two regions bargained with each other to oppose legislation which otherwise might have been enacted. For a good general discussion of the fate of these reform initiatives, see Cohen (1984).

11. One suspects that this is a grossly unreported phenomenon, simply because no one is looking for it. It is most often noted at upper levels of hierarchial societies. See Schapera (1943). But it can also occur at village level (Bruce 1976).

12. A thorough going look at the literature on participation is provided by Uphoff, Cohen and Goldsmith, (1979). For a more direct application in the land reform context, see Montgomery (1972).

References

Adeyoju, S.K. 1975. Where forest resources improve agriculture. *Unasylva* 27:28-29.

Ampuero, E.P. 1979. Ecological aspects of agro-forestry in mountain zones: the Andean region. In T. Chandler and D. Spurgeon (eds). *International Cooperation in Agro-forestry: Proceedings of an International Conference.* DSE and ICRAF. Nairobi.

Barraclough, S. and A. Domike. 1966. Agrarian structure in seven Latin American countries. *Land Economics* 42:391-424.

Bernard, F.E. and D.J. Thom. 1981. Population pressure and human carrying capacity in selected locations of Machakos and Kitui districts. *Journal of Developing Areas* 15:386-387.

Berry, S.S. 1975. *Cocoa, Custom and Socio-Economic Change in Rural West Nigeria.* Clarendon Press. Oxford.

Bloch, P. 1985. Land tenure issues in river basin development in Sub-Saharan Africa. Land Tenure Center. Madison, Wisconsin.

Boonkird, S.A. 1979. Taungya system: It's applications, ways and means of improvements in Thailand. In *Proceedings, Vll World Forestry Congress.* IUFRO. Jakarta.

Boutilier, J-L. 1963. Les rapports du systeme foncier Toucouleur et de l'organisation sociale et economique traditionelle-leur evolution actuelle. In D. Biebuyck (ed). *African Agrarian Systems.* International African Institute. Oxford University Press. London.

Brain, J. 1980. The Uluguru land usage scheme: success and failure. *Journal of Developing Areas*: 175-190.

Bruce, J. 1976. Land reform planning and indigenous tenures: A case study of the Chiguraf-Gwoses in Tigray, Ethiopia. S.J.D. dissertation. University of Wisconsin.

Bruce, J. 1981. Observations on land tenure and housing development in the major villages of Botswana. LTC Research Paper No. 75. Land Tenure Center. Madison, Wisconsin.

Bruce, J. 1983. Family land tenure and agricultural development in St. Lucia. LTC Research Paper No. 79. Land Tenure Center. Madison, Wisconsin.

Carneiro, R. 1971. Slash-and-burn cultivation among the Kikuru and its implications for cultural development in the Amazon Basin. In Y. Cohen (ed). *Man in Adaptation: The Cultural Present.* Aldine-Atherton. Chicago.

Cohen, J. 1984. Foreign involvement in land tenure reform: The case of Ethiopia. In J.D. Montgomery (ed). *International Dimensions of Land Reform.* Westview Press. Boulder, Colorado.

Coulson, J.J. 1971. *Succession in the Muslim Family* Cambridge University Press. Cambridge.

Denevan, W.M.. 1982. Causes of deforestation and forest and woodland degradation in tropical Latin America. Report to the Office of Technology Assessment, Congress of the United States. Assessment of Technologies to Sustain Tropical Forest and Woodland Resources. U.S. Office of Technology Assessment. Washington, D.C.

Deng, L.A.L. 1984. The Abyei development project: A case study of cattle herders in the Sudan. Ph.D. dissertation. University of Wisconsin.

Duncan, P. 1960. *Sotho Laws and Customs.* Oxford University Press. Cape Town.

Eckholm, E. 1967. *Losing Ground: Environmental Stress and World Food Prospects.* W.W. Norton and Co. New York.

Eckert, J. 1980. *Lesotho's Land Tenure: An Analysis of Annotated Bibliography.* Special Bibliography No. 2. Ministry of Agriculture, Kingdom of Lesotho and Department of Economics, Colorado State University.

Foster, G.M. 1966. *A Primitive Mexican Economy.* Monographs of the American Ethnographical Society No. 5. University of Washington Press. Seattle.

Gluckman, M. 1965. *Politics, Law and Ritual in Tribal Society.* Basil Blackwell. Oxford.

Goswani, P.C. 1982. Agro-forestry practices and prospects as a combined land use system. *Indian Forester* 108:385-396.

Hammer, T. 1983. Socio-economic implications of fuelwood development in poor countries. DERAP Working Paper No. A288. Chr. Michelsen Institute. Bergen, Norway.

Harbeson, J.W. et al. 1984. *Area Development in Liberia: Toward Integration and Participation.* AID Project Impact Evaluation No. 53. AID. Washington, D.C.

Hill, P. 1963. *Migrant Cocoa Farmers of Southern Ghana.* Cambridge University Press. Cambridge.

James, R.W. 1967. Judicial developments in customary law: Some reflections. *Journal of the Denning Law Society* 2:41-58.

James, R.W. 1971. *Tenure and Policy in Tanzania.* East African Literature Bureau. Nairobi.

James, R.W. and G.M. Fimbi. 1973. *Customary Land Law of Tanzania.* East African Literature Bureau. Dar es Salaam.

Jones, J. 1982. Sociocultural restraints in working with small farmers in forestry: The case of land tenure in Honduras. In *Short Course in Agro-forestry in the Humid Tropics.* INFORAT. Turrialba, Costa Rica.

Land Tenure Center. 1982. Project identification document: Land tenure, Republic of Liberia. Land Tenure Center. Madison, Wisconsin.

Land Tenure Center. 1985. Research project paper: changes in agricultural land use: Institutional constraints and opportunities. Land Tenure Center. Madison, Wisconsin.

Macfarlane, A. 1976. *Resources and Population.* Cambridge University Press. Cambridge.

Meek, C.K. 1949. *Land Law and Custom in the Colonies.* Frank Cass. London.

Meggers, B. 1971. *Amazonia: Man and Culture in a Counterfeit Paradise.* Aldine-Atherton. Chicago.

Millon, R.F. 1955. Trade, tree cultivation, and the development of private property in land. *American Anthropologist* 57:698-712.

Montgomery, J. 1972. Allocation of authority in land reform programs: a comparative study of administrative processes and outputs. *Administrative Science Quarterly* 17:62-80.

Morgan, W.B. 1969. Peasant agriculture in tropical Africa. In M.F. Thomas and G.W. Wittington (eds). *Environment and Land Use in Africa.* Methuen. London.

Nadel, S.F. 1942. *A Black Byzantium.* International African Institute. Oxford University Press. London.

Nadjombe, O. 1981. Taungya practices in Togo. In L.H. MacDonald (ed). *Agro-forestry in the African Humid Tropics.* United Nations University Press. Tokyo.

Noronha, R. 1981. Why is it so difficult to grow firewood? *Unasylva* 33:4-12.

Ogundowara, O. and E.O. Heady. 1973. Integrating short-term farm enterprises with perennial tree crops: an application of recursive programming to a tree crop settlement in Western Nigeria. *Nigerian Journal of Economic and Social Studies* 73:81-94.

Okeyo-Pala, A. 1978. Women's access to land and their role in agriculture and decision making on the farm: experiences of the Joluo of Kenya. Institute of Development Studies. Discussion Paper No. 263. IDS. Nairobi.

Rald, J. and K. Rald. 1975. *Rural Organization in Burkoba District.* Institute of African Studies. Uppsala.

Renard, A.S. and R.D. Renard. 1980. Introducing the resource system of development, with special reference to the Karens of northern Thailand. In *Report of the FAO/SIDA Seminar on Forestry in Rural Community Development.* FAO. Rome.

Rimoldi, M. 1966. Land tenure and land use among the Mount Lamington Orokaiva. New Guinea Research Bulletin No. 11. Australian National University, Canberra.

Schapera, I. 1943. *Tribal Legislation Among Tswana of the Bechuanaland and Protectorate.* Monographs on Social Anthropology No. 9. London School of Economics and Political Science. Percy Lund, Humphries and Co. London.

Seidman, R.B. 1968. Law and economic development in independent, English-speaking, Sub-Sahara Africa. In T.W. Hutchinson (ed). *Africa and Law.* University of Wisconsin Press. Madison.

Sellers, S. 1977. *The Relationship between Land Tenure and Agricultural Production in Tucurrique, Costa Rica.* CATIE. Turrialba, Costa Rica.

Shambi, M.M. 1972. *The Problem of Land Ownership and Cashewnut Claims in Malindi Coastal Belt.* Kilifi.

Soetrianegara, I. 1982. Socio-economic aspects of forest resource management in Indonesia. In F.G. Hallsworth (ed). *Socio-Economic Efforts and Constraints in Tropical Forest Management.* Wiley. Chichester.

Sorbo, G.M. 1980. Off-scheme interests and economic differentiation in Sudanese tenant communities. Paper presented at a Sociology Seminar. World Bank.

Sossah, K. 1980. *Migration Provoguee a Bassa.* Societe Togolaise d'Etudes et de Developpement (SOTED). Lome, Togo.

Suganob, R. 1980. Reforestation problems in the province of Cebu. *Canopy* 6:45.

Tanner, R. 1960. Land rights on the Tanganyika coast. *Africa Studies* 19:14-25.

Thomson, J.T. 1982. *Participation, Local Organization, Land and Tree Tenure: Future Directions for Sahelian Forestry.* Club du Sahel/OECD.

Tiffen, M. 1984. Land tenure issues in irrigation planning design and management in Sub-Sahara Africa. Utah State University Water Management Synthesis Project. Logan, Utah.

Tiffen, M. forthcoming. *Economic, Social and Institutional Aspects of Shifting Cultivation on Humid and Semi-Humid Africa.* FAO. Rome.

United Nations Development Program (UNDP). 1980. *Rural Women's Participation in Development.* Evaluation Study No. 3. UNDP. New York.

Uphoff, N., J. Cohen, and A. Goldsmith. 1979. Feasibility and application of rural development participation: a state of the arts paper. Rural Development Committee. Cornell University. Ithaca.

USAID/Mauritania. 1981. Project paper: Mauritania Land Tenure Project. USAID. Nouakchott, Mauritania.

Wilson, M. 1963. Effects on the Xhosa and Nyakyusa of scarcity of land. In D. Biebuyck (ed). *African Agrarian Systems.* International African Institute. Oxford University Press. London.

Wong, D. 1975. *Tenure and Land Dealings in the Malay States.* University of Singapore Press. Singapore.

Yengoyan, A.A. 1964. Environment, shifting cultivation and social organization among the Mandaya of Eastern Mindanao, Philippines. Ph.D. dissertation. University of Chicago. Chicago.

Zeidenstein, S. (ed). 1979. Learning about rural women. *Studies in Family Planning 10.*

Part 2

Regional Position Papers

Land Tenure and Agroforestry Land Use Systems in Ghana

George Benneh

University of Ghana, Legon

Introduction

In the last few years, agroforestry has acquired a reputation as an applied multidisciplinary science that can provide solutions to some of the major crises which many of the Third World countries face. Crises such as the food crisis, the "other energy crisis," creeping desertification and land degradation, it is argued, can be resolved through the application of agroforestry techniques. Indeed, there is the danger that too much may be claimed for agroforestry too soon, especially if one realizes that although agroforestry is a new term the practice of resource management which includes trees and crops is certainly not new to farmers in many Third World countries.

This is not to underestimate the contribution that agroforestry can make toward the achievement of sustainable development in many countries, if recommended techniques and practices are based solidly on adequate knowledge of existing systems of land use and cultural practices.

One of the important components of any land use or farming system is the land tenure system. The institutional arrangements under which a person gains access to land largely determines, among other things, what crops he can grow, how long he can till a particular piece of land, his rights over the fruits of his labor and his ability to undertake long term improvements on the land. It may thus be said that every tenure system is closely associated with a characteristic land use system. A change in a land use system, therefore, is likely to lead to a change in the tenure system, and *vice versa*.

In introducing a new type of land use, it is necessary to ascertain whether the recommended land use can be adopted within the framework of the existing land tenure system. If it cannot, it will be necessary to determine the changes which should be effected in the land tenure system to accommodate the new land use system.

The need for field surveys to elicit this kind of information cannot be over-emphasized, since in a dynamic situation what has been handed down as customary practice may often run counter to contemporary social reality.

This paper will deal briefly with the following issues: 1) traditional agroforestry land use systems in Ghana; 2) the relationship between land tenure and these land use systems; 3) the extent to which these land use systems are stable in the light of the growing population pressure on available cultivable land and the demand for fuelwood for domestic use; and 4) a review of the introduced or proposed agroforestry land use systems and the implications they have for tenure systems in the country.

Traditional agroforestry land use systems

Although agroforestry has many definitions few would dispute the fact that it is a multiple land use system involving the combination of forest trees with crops or with domestic animals or both. Trees are usually planted or at least tended for their products or their services (Budowski 1982).

Based on the above definition two major agroforestry land use systems may be identified. These are the bush fallow system and cash tree cropping (see Benneh 1973 for a more detailed discussion of these systems).

Bush fallow system

The bush fallow system is the dominant peasant system of farming in both the forest and savannah regions of Ghana. The main characteristics of the system are rotation of fields rather than of crops, easy acquisition of land for cultivation, use of fire for the clearing of vegetation, dependence on muscle power, use of simple implements for cultivation, mixed cropping and the restoration of the fertility of the soil by fallow vegetation. Although some of the trees may be cut down and burned or used as fuelwood, others are left standing on the plot to provide shade for certain cultivated crops.

A bush fallow plot is normally fringed by fallow vegetation in different stages of growth from which other household subsistence needs--such as medicinal plants, fuelwood, chew sticks and ropes--can be obtained.

A bush fallow plot in the forest region simulates a tropical forest in its degree of generalization. It is characterized by the intercropping of many types of domesticated plants. The description of the system as one in which a natural forest is transformed into a harvestable forest cannot be more apt.

In the northern savannah regions of Ghana economically useful trees particularly *Butyrospernum paradoxum* sub. sp. *parkii* (shea nut), *Acacia albida*, *Adansonia digitata* (baobab), *Parkia clappertoniana* (dawadawa), and others are integrated with food crops. In these areas, especially where there is pressure of population on land, only these economic trees are left. They are not planted and most of the trees are very old. No provision is made for the planting of fuelwood and fodder trees.

Under the bush fallow system, that member of the land owning group who is the first to cultivate a piece of unclaimed land establishes for himself and his descendants the right to use that land for cultivation. His rights extend to the fallow vegetation. Boundaries of individual farm lands are not well defined but they can be determined in times of dispute.

Since farmers are aware of differences in soil and their suitability for the crops they grow, each farmer tries to establish a number of fields at different sites to take advantage of the available soil types. This leads to fragmentation of holdings.

At any particular site the fields of an individual may be interspersed with those of others giving rise to a mosaic pattern of landholdings with no characteristic shape of fields. This general pattern differs from the strip pattern of ownership known as *huza* found in the eastern region of Ghana. A variation of the bush fallow land use system is the combined field system, where a farmer cultivates intensively a permanent field around his compound in addition to a bush fallow farm at some distance from the settlement (Benneh 1973).

Cash tree cropping

Cash tree cropping became important in Ghana with the successful introduction of cocoa in 1879. Its rapid expansion in the forest zone was partly due to the fact that its adoption did not necessitate a break with the traditional bush fallow system.

The land for tree cultivation, whether of cocoa, oil palm, coffee or rubber, is prepared in much the same way as that for food crops under the bush fallow system. Seedlings of the tree crop, which are raised on beds at suitable sites, are transplanted to the prepared field. Shade for the young seedlings is provided by a few trees left standing on the cultivated plot and by food crops such as banana, plantain and taro. Some trees are favored as shade trees. These include *Ceiba pentandra*, *Chlorophora excelsa* and *Ficus asperifolia*.

Access to land for tree cropping is gained by a member of the land owning group in much the same way as that for food cropping under the bush fallow system. The major change in land tenure is in respect of strangers (non-members of the land owning group) whose access to land is gained through direct purchase, tenancy arrangements or through gift (Benneh 1984).

Tension points in the traditional agroforestry land use system

The bush fallow system will function efficiently and be stable only if there is enough land to ensure that each abandoned clearing is allowed an adequate period of rest and regeneration. Gourou (1966) considered that a minimum of 25 to 30 years fallow is necessary for the soil to build its lost fertility. A more practical period of fallow has been suggested by Morgan and Pugh (1969). Field inquiries indicate that in the more densely settled areas, near major centres of population in the country, the fallow period is 2 to 3 years.

The effect of this reduction in the length of fallow has been the disappearance of tree species in the fallow vegetation, a process which has been exacerbated by the increased demands of the population for fuelwood. Yields of crops have declined and the cassava plant, which is tolerant of low soil fertility, has become the dominant feature of the agricultural landscape. In the northern savanna regions, frequent bush fires have not only changed the vegetation but caused erosion and land degradation.

Cash tree cropping is under a different kind of stress. It entails meeting the insatiable demand for new lands by migrant farmers who have exhausted the lands in their local areas suitable for cash tree farming. These farmers look upon forest reserves created by Government, and which cover about ten percent of the area of Ghana, as potentially the best lands for cash tree cultivation.

Although forest reserves have been established on lands acquired by Government, there are serious problems of encroachment by farmers with the connivance of traditional authorities who formerly owned these lands. There is clearly a conflict of land use between agriculture and forestry, not to mention differences in perception between farmers and Government in the use of a natural resource.

Meeting the present and future needs of the population through improved agroforestry practices and the land tenure factor

By far the most important agroforestry land use system introduced into the country is the *taungya* system. This has involved the release of land under forest reserves to food crop farmers in areas where there is land hunger. It is popular with the landless who find this the only way of obtaining access to land in their areas. The main difficulty with this system is in monitoring the activities of farmers so that they do not cultivate tree crops on the plots released to them in order to establish permanent claims over the land. As the *taungya* system is mainly practiced in the reserved forest lands, the main land tenure problem is that of encroachment.

Among the modern agroforestry land use systems which have been recommended for maintaining soil fertility and for energy production in Ghana the widely researched alley cropping system, the cultivation of multipurpose tree fallows, and the establishment of community fuelwood plantations.

The first two of these systems may be introduced without any major changes in the customary tenure relations of members of land owning groups. However, in the case of very short-term tenancy arrangements there will be a need for some modification in the terms of the contract for the tenant in order for him to benefit from the improved condition of the soil and the fruits of his labor. Again, in some cases, tenants are prevented from growing tree crops on land leased to them.

The establishment of community fuelwood plantations outside the areas acquired by Government may also pose some difficulties, especially in the land hungry areas and in the area of the *huza* system in the eastern region since there are no communal lands. At best, such plantations may be established on badly degraded lands with the permission of their owners.

While the customary tenure system can adjust itself in response to innovations which are based on small farms, because holdings are typically small and fragmented, tenure adjustment becomes a major obstacle when the scale of operation changes.

References

Benneh, G. 1973. Small scale farming systems in Ghana. *Africa* 43(11): 134-146.

Benneh, G. 1984. Dynamics of land tenure and agrarian systems in Africa: Ghana case study. Unpublished paper commissioned by FAO. Rome.

Budowski, G. 1982. The socio-economic effects of forest management on the lives of people living in the area: The case of Central America and some Caribbean countries. In E.G. Hallsworth (ed). *Socio-economic Effects and Constraints in Tropical Forest Management.* Wiley. Chichester.

Gourou, P. 1966. *The Tropical World.* Wiley. New York.

Morgan, W.B. and J.C. Pugh. 1969. *West Africa.* Methuen. London.

Tenurial Considerations in Agroforestry in Nigeria

S. Kolade Adeyoju

Department of Forest Resources Management, University of Ibadan, Nigeria

Introduction

Modern forestry is basically an enterprise which is alien to tropical economies. It has had to compete with shifting cultivation in various ways. Large tracts of relatively undisturbed land were reserved to forest use by preempting the development of individual rights, while the ownership of occupied forests was still being contested. During the reservation phase forty to fifty years ago, it was rather easy to abrogate the customary rights to land, as long as the chief or community leader was satisfied that the objectives of the proposed project were worthy of support. In those days, the population was much smaller than today, and the chief's pronouncements enjoyed as much sanctity as present day government edicts and decrees. Other factors have also come into play over the last few decades. There has been a rapid increase in population, while the cultivated lands have suffered nutrient depletion and erosion due to shorter cropping cycles. The sites previously left fallow for long periods now have to be cultivated more frequently by a considerably larger number of farmers. Moreover, the European concept of "freehold" tenure has been embraced over the years by a growing number of educated people who are genuinely dissatisfied, both with the authoritarian status of chiefs and the nebulous communal ownership of lands. In the circumstances, therefore, the existence of forest reserves has compounded these socioeconomic changes.

Ten percent of the area of Nigeria is gazetted forest, thus conferring on the forest services the responsibility of being the largest and most dominant estate agents in the country. Because forest reservation was carried out in the greatest haste, designed to beat the onslaught of shifting cultivators, and because there has never been a land capability survey, the foresters had to lay claim to large tracts of land, which embrace all types of vegetation, soils, and ecological zones. At that time, the reservation doctrine was that 25 per cent of the country had to be reserved for the supply of forest goods and services to the population. This was a useful guideline, without which much of the present forest cover could not have been salvaged. Although less than half the national target of 25 per cent forest land was reserved, the best agricultural lands are in some places under the auspicies of the forest service, and in certain districts, there is real agricultural land hunger. This is why there has been a sustained demand for dereservation. Indeed, in the eastern states of Nigeria not less than 2,082 square miles of forest reserve or potential forest reserve were lost after gazettement in one form or another (Rosevear 1953). Also in the western states, governments have had to approve excisions from forest reserves to ease proven land shortages.

However, in other places, the call for dereservation rests on the fact that although the agricultural land is sufficiently extensive to support more than the expected 20 to 30 farmers per square mile (Nye and Greenland 1960) the land has been misused. Because no soil conservation measures have been practiced, vast areas are now unsuitable for sedentary agriculture. Naturally, the first area to be attacked in the search for new land is the forest estate. This is the situation in most parts of the southern states where, superficially, there appears to be land shortage. However, with the flight of youth from the rural areas and an aging agricultural population, there should be a substantial increase in the total cultivable land area per farmer. Consequently, the reasons for dereservation in most places are related less to land shortage than to "forest-hunger."

Agricultural land shortage and forest land tenure

With a population of about 80 million, Nigeria has an average density in excess of 180 persons per square mile. It has therefore reached the danger level and the fear of a larger-scale food shortage is no longer a forecast but a reality. Although the drought problem has intensified food shortage, the basic cause is, indeed, the worsening performance of the present agriculture. In approaching this problem during successive national development plans, the Federal and State Governments have always proposed a variety of integrated policy measures concentrated on the following: improved production techniques, seed multiplication schemes, improved storage, organized marketing, fertilizer supply and subsidy, and application of pesticides. While these are certainly good intentions, some of them are probably too general and vague and, in any event, are of little value as short-term remedies, given the current level of technological development and the continuing ineffective role of the agricultural extension services.

A more fundamental approach to the problem requires a review of the status of land; that is, its ownership, nutrient status, improvement potential, and maximum use capability. These enquiries may be simplified by viewing them against the background of a broad outline of the philosophies of the two main land users: the forester and the farmer. In Nigeria, the farmer produces many crops on a given piece of land. For instance, in a subsistence plot as many as nine different crops may be raised during the growing season. Similarly, within the cash crop plot, six or more products may be produced simultaneously, although most of these would be long gestation crops. A few perennial crops may also be raised in the income-oriented plot to supplement production from the subsistence plot. Essentially, therefore, the peasant farmer is a naive multiple land user.

In theory and practice, tropical forestry is largely a single use enterprise. It is only in recent times that the compatibility of wildlife management, recreation, and watershed protection with timber production has been accepted. Even where this combination is in vogue, it is strictly on a zonal or unit area basis, since it is hardly feasible to pursue these various production objectives simultaneously on the same piece of land. However, there is one exception. Certain agricultural crops can be raised in the early stages of plantation establishment without severe impairment of the forestry production objectives (King 1968). Therefore, the introduction of agroforestry is one way of increasing the productivity of the forest land as well as stimulating greater benefits from joint inputs from the farmer and the forester.

The tenurial implications of this scheme are of particular concern to the forester. Although agroforestry is, in fact, a distinct modification of the system of shifting cultivation, the affinities are more physical and biological than legal. An essential ingredient of the system of agroforestry does not exist in shifting agriculture, that is, the land utilized under shifting agriculture is not owned and its development planned and controlled by a second party as it is in agroforestry. As Gordon (1955) has pointed out, shifting cultivation is generally not practised under a legal system which recognizes individual freehold rights. The cultivator establishes exclusive rights, often of a vague and unascertainable nature, through his clearing of family and tribal land, and these rights are generally temporary. On the other hand, in agroforestry, the land may be owned by the operator with technical advice coming from the Forest Department. In addition, the rights and duties of the farmer (operator) are often clearly defined; the regulations thereof become an important factor in land use. Thus the rights which the shifting cultivator formerly enjoyed through status are now his own as a result of contract.

The degree of contractual agreement may be clearly brought out in an investigation of the terms and conditions under which farmers are permitted to plant their crops in the forest reserves. In most states of Nigeria, a permit is issued formalising the arrangement. The permit confers on the farmer the right to use the forest land free of rent; to plant only specific farm crops in the area specially alloted to him; to keep his farm clean and tend the forestry and food crops together; and not to dispose of any forest produce. The permit is non-transferable and may be renewed for a second year. The situation in Bendel State is relatively more fluid: there have been frequent modifications of the terms and conditions of the agreement as a result of various abuses and allegations. Indeed, the system has been intermittently suspended and reactivated in Bendel over the last ten years.

In his discussion of the nature of agri-silviculture as a "licence," King (1968) observed that in many of the countries that base their law on the English legal system, a licence is regarded not as a separate entity, but merely as part of the contract creating it, so that even if the contract is not specifically enforceable the licensee is probably protected. Thus to take a hypothetical case, the courts may recognise the cultivator's right to enter upon forest land after he has moved to another area, so that he may harvest fruit from, say, his banana or cocoa trees which he had previously legally planted under an agri-silvicultural agreement. Consequently in the case of cocoa taungya in Ikom (Cross River State), the special terms of tenancy would be related to the cocoa regime. On the other hand, the Ijaiye cocoa farmers (Oyo State) who unilaterally moved into the forest reserve ten or twelve years ago are squatters.

While it is admitted that the presence of agriculturists in the forest reserves imposes restricted tenure conditions on the forester, the point should also not be overlooked that if faced with a growing food shortage from depleted and misused lands, society would have to make a choice between famine and respite. There can be no doubt about the rational decision that would be taken in that circumstance. Nevertheless, the obligation to keep the land under tree crops and maintain production, and the restrictions against subdivision of the land upon sale or inheritance, should be paramount in most situations. Also, the forest services that are failing to attract adequate investment for their resources might consider whether part of the reason does not lie in the rather inviolable nature of their tenures and give some thought to conditional freehold through agroforestry as one possibility that has not yet been fully investigated.

General considerations

The two major factors in favor of agroforestry tenure are a) flexibility and b) high investment potential (Adeyoju 1984). Flexibility as an objective in land tenure implies that opportunities exist for movement up the agricultural ladder. It means introducing a tenure system which creates opportunities for industrious laborers to become farmers and for ambitious customary tenant farmers to acquire absolute interests in land. More importantly, it means that whoever wants to practice mixed farming, including both specific forest crops and annuals, will have authorization and legal support to continue his enterprise until the forest crops are ready for harvest. Access to these opportunities should not be hindered.

Consequently, the policy that is envisaged should include provisions for such facilities as credit for capital formation, opportunity to market crops at reasonable prices, and the creation of adequate infrastructures. A system of land tenure that is flexible is never static. Rather, it should be dynamic, and change in accordance with the new social and economic features of the population for which it is designed.

The security of tenure in land encourages the farmer to make necessary investments. In other words, provision of security creates the opportunity for the farmer to raise the status of agriculture from a subsistence base to one oriented toward investment. Agroforestry is far more capital-demanding than traditional agriculture and thus would constitute a major investment in rural areas.

The most important social objective of land tenure policy is to facilitate the development of a well integrated rural community. The trends in agroforestry are strongly directed towards the development of prosperous rural communities, but again these efforts are localized within consolidated forest estates. Outside the forest estate, necessary incentives should be given to agroforestry farmers whose willing participation will promote economic integration.

References

Adeyoju, S.K. 1984. Some tenurial and legal aspects of agroforestry. In J.K. Jackson (ed). *Social Economic and Institutional Aspects of Agroforestry*. United Nations University. Tokyo.

Gordon, W.A. 1955. *The Law of Forestry*. Her Majesty's Stationery Office. London.

King, K.F.S. 1968. *Agri-Silviculture (The Taungya System)*. Bulletin No. 1. Department of Forestry, University of Ibadan. Ibadan.

Nye, P.E. and D.J. Greenland. 1960. *The Soil Under Shifting Cultivation*. Commonwealth Agricultural Bureau. Farnham Royal, England.

Rosevear, D.R. 1953. Forestry. In *The Nigeria Handbook*. Government Printer. Lagos.

Land Tenure Systems and the Adoption of Alley Farming in Southern Nigeria

Paul Francis

International Livestock Centre for Africa, Ibadan, Nigeria

Introduction

Since 1980 the International Livestock Centre for Africa's Humid Zone Program, based at Ibadan, Nigeria, has been undertaking research on sustainable farming systems integrating crops and livestock. The basis of this work is fodder production through alley farming--the cultivation of arable crops between rows of leguminous forage trees. The tree species used have been *Leucaena leucocephala* and *Gliricidia sepium*, these being fast growing and capable of providing high quality fodder for small ruminants or nitrogen rich mulch for the maintenance of soil fertility (Sumberg 1984). Most households in rural southern Nigeria keep sheep or goats. Flock size typically ranges from 2-6 animals.

ILCA's research has had an increasing on-farm component, and in 1983 on-farm trials were established at three sites, one in the southwest and two in the southeast of Nigeria. Farmers in villages at each site were encouraged to plant browse trees and were provided with seeds and advice about the planting and management of alley farms. The establishment and utilization of these farms have been monitored closely by ILCA staff. This paper is a preliminary review of the land tenure issues that have emerged from ILCA's on-farm experience to date.

The planting of browse trees represents a new use of land, and their utilization entails new patterns of activity. While the planting of tree crops is scarcely an innovation in itself in an area which has been a major producer of cocoa, kola and palm products, browse trees differ from these in that their product is not exchanged on the market but rather has a number of competing uses (as forage, mulch, staking material, firewood) as well as a number of potential beneficiaries (land owner, present cultivator, future cultivator, stock owner). Furthermore, alley farming, through its contribution to soil fertility, allows the possibility of extending the cropping period and thus of changing established land-use patterns and practices.

Here we consider several aspects of the land tenure systems at the three sites and their implications for the pattern and extent of browse tree adoption. These are presented in terms of a number of contrasts relating to the nature of rights to land on the part of women as compared with men, of natives as compared with immigrants, of landowners as compared with tenants, and of individual households as compared with corporate kin groups or the community.

Southwestern Nigeria

The southwestern pilot project is situated some 18 km northwest of Oyo in Oyo State in the neighboring villages of Owu Ile and Iwo Ate. Women account for 60 percent of the adult population of these two villages (which totals something over 500), but only 18 percent of the browse-planting participants. There are a number of reasons for the seemingly low level of interest on the part of women. In the first place, many women do not have farming as their primary occupation. Their main activities, in addition to their domestic work, are the processing and marketing of palm oil and cassava products and petty trading. Many of those who are directly involved in agricultural production work on their husbands' farms rather than on their own. According to a survey of the entire population of the two villages, only 29 percent of adult women farm and only 7 percent, over half of whom are widows, do so independently of their husbands (figures derived from Okali and Cassaday 1984).

Owu Ile and Iwo Ate contain a large number of people who consider themselves primarily members of other communities. These are by no means always recent immigrants, the category "strangers" is a heterogeneous one including as it does seasonal migrant laborers and those who have been resident in the community for over a generation but have maintained affiliations with another community such as the town of Oyo. It is thus somewhat misleading to treat all nonnatives together. Nevertheless, statistics from the demographic survey do not indicate that in general strangers found any particular difficulty in obtaining land for alley farming: those of the adult male population who define themselves as strangers account for 46 percent of the population and 45 percent of the alley farming participants.

Southeastern Nigeria

The two project areas in the southeast of Nigeria are situated at Mgbakwu near Awka in Anambra State and Okwe near Umuahia in Imo State. Women in the southeast of the country have traditionally been much more involved in agriculture than those in the southwest and therefore a higher level of participation might have been expected at these sites. Nevertheless, of the 17 farmers who planted browse trees at the two sites in 1984 only two were women. While almost all women farm, few farm independently. According to the demographic surveys of Mgbakwu and Okwe, none of those women who head production units in the two villages have husbands who are engaged in agriculture. A woman can only be said to hold and manage land on behalf of her husband or, in the case of widows, his relatives or children. A woman must therefore seek permission to plant browse trees from her husband or his kin and this, it seems, is not readily granted.

With regard to the participation of tenants, while a considerable proportion of land is rented out both at Mgbakwu and Okwe, only one of the farms planted with browse trees was on rented land.

While differential access to land on the part of women as compared with men, and owners as compared with tenants, is of critical importance to the participation of particular groups there are more fundamental features of the land use and land tenure systems at the sites in the southeast which have a bearing on the acceptability of browse trees. These concern prevailing systems of land rotation and the related issue of the rights of lineage and community to manage and allocate land.

In both Mgbakwu and Okwe "compound farms" (*ani uno* or *ali ulo*) in the immediate vicinity of the house are distinguished from ordinary fields or "distant farms" (*ani agu* or *ala ubi*). The precise meaning of these terms seems to differ between the two communities, but broadly speaking compound land is either cropped continuously (at Mgbakwu) or with a very short fallow period (one year at Okwe) and is considered to be owned by the individual householder.

Outside the compound, the usual practice on the fragile soils of the southeast is to plant maize, cassava, and (in Mgbakwu) yam in the first year of cultivation and, having harvested the maize and yam at the end of the first season, to leave the cassava with minimal management and generally without weeding until it is harvested a year or more later. The land then reverts to fallow for a period which depends on land pressure in the area and distance of the land from the settlement, although the fallow period is typically some three years in Mgbakwu and six in Okwe. In the southwest of Nigeria, where land is cropped for two seasons a year for several years continuously before being allowed to revert to fallow, the presence of browse trees does not necessitate *extra* weeding in the second and subsequent years of cropping since the land remains under cultivation. Under the southeastern system, however, where land would normally be abandoned, any labor expended in the maintenance of alley farms is additional to normal requirements. As the neglected condition of many alley farms testifies, farmers have been reluctant to make this investment where the potential benefits are not clear. Conversely, the potential of alley farming to extend the cropping period through the application of prunings as mulch will not be realized unless alley farms are more intensively managed. Furthermore, the farmer's incentive to use the alley farming system to maintain soil fertility and allow continuous cropping is reduced by certain customary arrangements for the joint management of land, to which we now turn.

Mgbakwu consists of a group of six villages each of which in turn is comprised of a number of lineage segments (*umunna*, commonly translated as "families") who trace their descent from a common ancestor in the male line. These units control land, which is allocated by them annually to member households and which reverts to them at the end of a cropping cycle. Any land in excess of the families' requirements is rented out on a short term basis to tenants, who are usually members of other local *umunna* whose own land is insufficient.

At Okwe, the unit of ownership of land is the individual, rather than the lineage, but common control over its exploitation is in part vested in a residential group, the village (there are six constituent villages in the Okwe project area). The boundaries of village lands are well defined and it is said to be jointly decided by senior members of the village which sector of land is to be worked in any year. Villagers who do not hold land in this sector, even if they hold land elsewhere are expected to hire from those who do. Oil palms in Okwe are also jointly managed by the village, days being set on which those subscribing to a village-based fund may cut palm fruit from the communal trees.

For the present purposes, these two rather different systems for the control and allocation of land both tend to reinforce the seeming reluctance to engage in alley farming noted above. Under the Okwe system, although individuals own land, they are regularly compelled to rent other land by the system of joint management just described. While involved in temporary (and somewhat expensive) lease arrangements they will have little incentive to devote their labor to improving the future fertility of the land. Under the Mgbakwu system both tenants and members of land allocating families hold only temporary usufruct rights over land, and there would be a similar lack of incentive to invest in soil fertility.

Furthermore, under both the Okwe and the Mgbakwu systems of tenure, the extension of the cropping cycle by individuals would throw them out of phase with the pattern of rotation set by the village or the *umunna*, respectively, and thus lay them open both to the censure of other members of these groups and to the greater risk of animal pests attacking their crops when farming in isolation.

Conclusion

The above account of the relationship between browse tree adoption and land tenure systems is a preliminary one but suggests some tentative conclusions as well as the need for further research in a number of areas. It is clear that the question of "land tenure" cannot be separated from, on the one hand, the structure of authority within the household and, on the other hand, established patterns of cropping and rotation. At the southwestern site, it would seem that customary occupational roles and the prevailing structure of economic opportunity rather than land tenure rules as such explain the low participation of women. In the southeast, the apparently low involvement of women is an aspect of the structure of decision making within the household production unit. Corporate control over land in the southeast seems to have been reinforced by the pressure of population, and this ironically makes the changes necessary to allow longer cropping cycles without soil degradation at the same time more urgent and more difficult to bring about.

To test these hypotheses, further research will be required on the nature of land rights of women and of the various categories of immigrants and on the extent to which even native farmers in the southeast are forced to rent land. Again, while it is clear that communal management and allocation practices influence cropping patterns, more needs to be known about the basis and effectiveness of communal decision making, including the sanctions at the disposal of the community. Lastly there is the question of the relationship between soil fertility, rotation, and land-use patterns. This will require close investigation of *actual* (as opposed to *stated*) land-use practices as well as of the indigenous soil classifications and concepts of soil fertility which underlie them.

References

Okali, C. and K. Cassaday. 1984. Women's participation in a livestock development programme. ILCA. Ibadan

Sumberg, J.E. 1984. Small ruminant feed production in a farming system context. ILCA. Ibadan.

Possibilities of Introducing Agroforestry in Zaire

Kalambay Lumpungu

Faculty of Law, University of Kinshasa

Translated from the French by Richard Labelle

Introduction

According to the terms of the land tenure legislation in effect since July 20, 1973, the state of Zaire is the sole land owner and each and every person under public or personal law can only have on the land a right of permanent or temporary enjoyment, called a perpetual concession or an ordinary concession of 25 years, with the possibility of renewal.

As a result of this law, all rights of land ownership acquired previously, have been converted into either a perpetual concession or into an ordinary concession. As regards lands of traditional communities, the law states that the president of the Republic will make an ordinance recognizing collective and individual tenure rights of the traditional communities. While awaiting this order, traditional communities continue to enjoy this land as in the past.

The control of all of the land by the state of Zaire has an important advantage in the ability to impose on all legal title holders new obligations, such as specific development requirements in relation to the dictates of town planning or of the environment.

Concerning land ownership and the adoption of agroforestry innovations

Studies on traditional tenurial rights undertaken by ethnologists as well as by legal writers toward the end of the colonial period have all come to the same conclusion: despite the size of Zaire and its large number of tribes, traditional land tenure is everywhere based on the same fundamental principles.

Each traditional community holds the land in a collective ownership that is well defined and recognized by neighbors. Each such community is composed of related groups, that is, people sharing or claiming to share a common ancestry. These people are, thus, the collective owners of community land. One could even say that they are co-proprietors.

If the land belongs to the community, the community may allow strangers to live there and exploit it. Membership in the group may be gained either by allegiance or by marriage; however, it must be noted that the rights of strangers on the land estate do not have the same character as those of natives. As just mentioned, members of the group are co-owners of the land, while strangers acquire usufruct rights of beneficial usage and that these rights cannot be inherited by their children, except with the previous authorization of the community chief.

The acquisition of individual rights of land ownership originate in the act of developing the land, either by building a house or through annual or perennial cropping. As long as the land is in use, individual rights are guaranteed. Once the land is no longer in use it reverts to collective ownership. The principles are valid both for community members and for strangers.

Shifting cultivation is the traditional agricultural practice. When soil fertility is exhausted, the farmer moves to another plot within the community land, thus establishing a fallow rotation within the area. Unfortunately, this cropping system is always accompanied on the one hand by much deforestation, which is harmful to the soil, and on the other hand by a displacement of the villages. Despite this, planted trees remain the property of those who planted them. Agroforestry may have a role to play in sedentarizing the population. Because of the perennial nature of agroforestry cropping systems, if a land user develops an agroforestry system on land under the co-ownership of the community, this may result in the developer becoming, in effect, the sole owner of the land.

This situation cannot overturn the principles of customary land tenure. However, for strangers, for whom customary law has always recognized a right analogous to usufruct that is personal, and thus for life, the question does arise of compensation of heirs for improvements on the land in the event of the death of the usufruct right holder, if the community chief does not wish the heirs to remain on community land.

Agroforestry permits the stabilization of the population, thus making it possible to open up a portion of the land that the state can give away to third parties. The concession of rural lands is generally an ordinary concession for a term of twenty five years. At the end of the term, taking population growth into consideration, the state can either refuse the renewal of the concession (lease), or agree to renew the lease on part of the land.

The stabilization of the population will simplify enormously the inquiry that is preliminary to all concessions of rural land. This constitutes a savings of time for the requester seeking a concession of rural land. The stabilization of populations due to agroforestry will also permit the state to establish an occupation map and to conceive, at the same time, a land tenure policy for leasing concessions and for land use management, and thus to realize, in the long term, integrated rural development, first on a regional then later on a national scale.

Agroforestry systems to solve or mitigate existing problems of land users

In 1936 the Belgian Congo inaugurated a rural development policy under the name of agricultural "paysannat" or "lotissement" (parceling out). This system is analogous to agroforestry, but different in that the peasant produced, on the same parcel of land, food crops and export crops such as peanut, cotton, maize and cassava. Paysannat is a form of rural economic organization in which, with the help of public institutions, native kinship groups develop and exploit for themselves land on which they will establish themselves permanently.

Schematically, we can say that the Congolese paysannat had three essential aims: 1) agricultural: the cropping system practiced in the paysannats conserved soil fertility through the adoption of adequate fallow systems in order to guarantee continuous production on the same unit of land; 2) social: the paysannat permitted the stabilization of the rural population and the improvement of the moral and material conditions of life; 3) economic: it increased production of foodstuffs to satisfy internal and export demand.

In the paysannat system, in order to receive title to a plot of land, the peasant had to fell the trees. This caused local microclimatic changes. Since agroforestry encourages the growing of trees together with crops, the paysannat system could be improved through the inclusion of multipurpose trees for soil fertility and erosion control in upland cropping systems. Zaire, however, is a vast and varied territory and one cannot prescribe one agroforestry system for the whole of it. A preliminary study is indispensable in order to take into account the land capability of the area in question.

In order to convince the farmer to try agroforestry, it would be necessary to use only fast growing multipurpose trees that must first be demonstrated to him. The farmer needs quick results to take up a new system. In localities that have not known the paysannat system, the results of agroforestry could be similar to those when plots are allocated, i.e. stabilization of the farmers plot, improved crop production and soil fertility, and reduction of erosion. By sedentarizing the population, it will be easier for the government to recognize individual land use rights as stipulated in the tenure law which demands beneficial occupancy and individual exploitation.

Disadvantaged groups

As noted above, the rights of outsiders on the land developed by them differ from those of members of the local group only in regard to rights of transfer between living individuals or inheritance upon the death of the original developer of the land. In other words, while alive the rights of outsiders are identical to those of members of the local group. Inheritance of these rights, as previously noted, is subject to the approval of the local chief.

In the event that the land developer is without heirs that are acceptable to the group, tradition is not clear on how these heirs are to be indemnified. We believe that this question will no longer be an issue once legislation is enacted recognizing individual land tenure rights on traditional lands. In any case, outsiders generally are able to acquire full rights by creating family ties with the local group, either directly through marriage or indirectly through the marriage of their children. In which case, problems of inheritance do not arise.

Unmarried females in the local group enjoy the same rights as the males as regards acquisition of tenurial rights by developing the land; similarly for widows from outside the group, especially those with children.

The Bantu conception excludes land from commercial exchange. Thus land cannot be rented, which is why customary law does not include tenant farming. Thus, as stated above, customary law only recognizes one type of right, corresponding to usufruct with certain restrictions.

Research needed to enlighten individuals concerning agroforestry

Agroforestry is a new science driven by a certain philosophy. Agroforestry cannot lead to the results expected without popular participation. Thus, before introducing agroforestry, certain studies are needed.

First, research must bear on customary political and administrative institutions that are within the traditional group, because these institutions persist independently of the political will to eliminate them in favor of the state.

Secondly, research on the tenurial systems of traditional groups is needed. This is important even in Zaire, where the principles are similar, since the particularities of each group must be recognized and accounted for.

Pedology is the third research objective. In this case, crops can be selected for the best yield on most appropriate land. This research should precede the introduction of agroforestry. Given that agroforestry is to be practiced by the farmer, its success will depend on the extent to which information on the advantages of agroforestry over traditional land use can be given.

In order to promote agroforestry among the people, demonstrations with group members must first be undertaken. This will permit comparisons to be made and the differences to be evaluated. For this to be successful, individuals must be trained. For Zaire, agroforestry should be introduced in the programs of the Institut Superieur des Sciences Agronomiques and the Institut Superieur du Developpement Rural. Graduates from both institutions could serve as extension agents.

Privatization of Land and Tree Planting in Mbeere, Kenya

David Brokensha and Bernard W. Riley

Institute of Development Anthropology and University of California, Santa Barbara

Introduction

This paper is a further consideration of issues dealt with in an earlier report (Brokensha et al. 1983) in which we examined a broad spectrum of factors associated with fuelwood use and deforestation. Here, we are specifically concerned with why people do, or do not, plant trees, using our fieldwork in Mbeere, Embu District as a case study.[1] We demonstrated earlier that individual land title (Brokensha and Glazier 1973), which is now universal in Mbeere, was a key factor in examining tree planting. However, it is by no means the only factor, nor is there a simple relationship. The granting of individual land titles does generally lead to an intensification of agriculture, which often involves the planting of trees. Long-term planning, essential for tree planting, is facilitated with the security of tenure which the titles provide. But many other factors must be considered in determining tree planting rates, types of trees planted, who plants, etc.

Most (not all) of the trees that are being planted in Mbeere can be legitimately considered as an aspect of agroforestry, in that they form part of the total farming system.

Environmental factors

Clearly, physical factors of elevation, soil and climate (especially the intensity and duration of rainfall) are of great significance. For our study, we divided Mbeere into three zones, which vary in elevation, rainfall, soil types, vegetation and fertility. Zone I, the highest, has an average annual rainfall of 1100 mm; Zone III, which goes down to the River Tana, receives less than 600 mm annual rainfall, so can be classified as "semi-arid", with all the associated problems of the ASAL (arid and semi-arid lands). Our 1982 survey[2] indicated that more trees were planted, per household, in Zones I and II than in Zone III. The mean numbers were, respectively, 60,62 and 23. It is hardly surprising that people in the more favored areas plant more trees than do those who live in harsher environments, where poor soils and low and unreliable rainfall are not conducive to the growth and survival of trees.

Social factors

Income and wealth

One of the most significant findings of our long term study is the growth of rural inequality, a fact that has been reported in detail from many other parts of the continent, and indeed from most developing countries. Each of our student interviewers selected, under our guidance, a "rich", an "average" and a "poor" household[3] in his or her home area, then completed a wealth inventory, which confirmed the subjective selections, as this part of the survey shows:

Item	Percentage of Households Owning		
	"Rich"	"Average"	"Poor"
Vono bed	68	53	48
Axe	94	85	81
Kerosene lamp	96	87	41
Watch or clock	94	70	10
Iron roof	98	78	4
Water tank	80	17	0
Bicycle	76	70	18
Wheelbarrow	71	39	2
Ox cart	35	15	0
Planted trees	74	52	14

Not only were fewer trees planted by the poorer people, but fewer people within this group had planted any trees at all.

We should add that we realize that the very poorest have been omitted from this short survey, confirming Robert Chambers' statement about the invisibility of the poorest of the poor. But even allowing for this, there are some clear differences in ownership of these selected items, and their significance is clear, especially when considering items that help increase farm productivity. Of course, off-farm income is a major factor in accounting for the differences.

Where people lived was of relatively less importantance than other factors in looking at ownership of these items (the differences between the three zones was less sharp), though there was a general diminution in percentage, from Zone I to Zone III.

Education.

This is strongly correlated with income, but not on a one-to-one basis. Some Mbeere farmers are quite wealthy in land and livestock, while having little or no formal education. The recent dramatic expansion of both primary and secondary schools in Mbeere has produced an important new class of rural people, young people who have had several years of education, often including some secondary schooling, and who are now engaged in farming. This class has been frequently overlooked, and we still read the old stereotypes about schooling producing a distaste for manual labor. Such people (often both husband and wife share this advantage) tend to be more receptive to advice and suggestions, and to be generally more innovative. Literacy is also a factor, especially for the thousands of Mbeere women who have successfully completed literacy classes, and whose confidence has, as a result, increased. They too tend to be more prepared to consider new ideas, and to have more confidence in themselves.

Religion

When we first visited Mbeere in 1970, we could easily pick out the households of prominent members of the Anglican Church, in the Mavuria and Nthawa locations, by their neat households which were distinguished by the presence of trees. Often there would be some shade trees and a small avenue leading to the house, as well as trees grown for more practical reasons. Even today, church groups are often active in promoting the planting of trees (see below).

Models

When one person leads, others tend to follow. In walking or driving around the area we have noted clusters of households that are actively engaged in tree planting, perhaps influenced and encouraged by a local extension officer, teacher or other change agent.

Idiosyncratic factors

We include this category because we have so many times been struck by the relevance of such factors. For example, some persons have the land and other resources to plant trees, but they never get around to it, partly because of their inability to organize themselves; others, with few advantages, seem able to plan and organize themselves so as to make impressive use of their limited resources. This is true in all societies, of course, so we were not surprised to find this phenomenon in Mbeere.

External factors

Government

Government has an important role in many ways, especially in providing effective extension officers who can reach many farmers (not only the richer ones and those who live near the main roads) and who can give them sound advice on tree planting and seedling availability. We have found that many people in Mbeere, as elsewhere in Africa, have never had occasion to plant a tree. They have used wood products for most of their material culture and have an impressive knowledge of the qualities of their woody vegetation, but until quite recently it was seldom considered necessary to plant a tree. There were thought to be enough available "in the bush." Now the communal bush has effectively disappeared, with the granting of individual titles, and trees (such as are left) are no longer freely available to all. The point of this is that in some cases, people have to be taught about the proper care of trees, as it is often a new thing for them.

Government should also provide nurseries, especially smaller decentralised ones, as people obtain more seedlings to plant when they are readily available, from a nursery in the vicinity.

Government has also been active, in Kenya, in what can be called "consciousness raising." There is constant repetition of the need to plant trees, through Presidential exhortations and those of other government officials, through radio messages and other media, and through the National Tree Planting Week. All of these efforts have had visible effects and should not be dismissed as mere rhetoric. People in rural Kenya *are* now more aware of the urgency of the task.

NGOs

Non-governmental organisations are playing major roles too, in some areas, in helping to establish nurseries and generally encouraging tree planting. Churches, women's groups and others are all active.

Private nurseries

Although, as far as we know, there are as yet no private nurseries operating in Mbeere, such commercial enterprises are common in the more densely populated areas of Kenya, and they also help promote the planting of trees.

Conclusion

All the indications are that the privatization of land in Mbeere has been a key factor in encouraging the planting of trees, because of the security it provides for long term planning. For example, the tree planting rate increased dramatically after people received their land titles; our survey shows that in the years 1980-1982, when people were beginning to feel secure in their new ownership of land, more trees were planted than in the whole decade of the 1970's.

We need to point out, though, that it is much easier for those who are not in the poorest economic levels to plant trees. The poorest may lack land altogether or have little land and feel that they must devote all of it to food or cash crops. Privatization of land has benefitted some, but has also been a constraint to others, such as those who no longer have access to the commons for their wood needs.

Trees are planted for many reasons, not necessarily those that officials consider the most pressing. Trees are planted for timber, for fruit and shade, and seldom explicitly for fuelwood. In Mbeere multipurpose trees are favored, as they are in many other areas. What is more important than the actual trees that are planted, is to encourage the habit of tree planting in the hope that it will spread to meeting other needs later.

Trees planted by individuals tend to have higher survival rates than do those planted communally, except for school plantations (providing the problem of school holidays can be overcome). There is a case for undertaking more specific, detailed local studies to explore further the relationship between individual title and tree planting, to consider the wider consequences of privatization, and to determine what sorts of assistance are needed (from both Government and NGOs) in order to accelerate tree planting. We end by noting that while dramatic progress has been made in Mbeere, there is a long way to go before the need for wood products can be met.

Notes

1. Mbeere now comprises Siakago and Gachoka Divisions. Our fieldwork was conducted from 1970 to 1983, with a total of nearly three years investigating social and ecological change with an emphasis on woody vegetation.

2. This survey was conducted with the help of nearly one hundred secondary school boys and girls, who obtained information on selected households in their home areas.

3. "Rich" and "poor" are, of course, relative terms, to be taken in the context of the local society.

References

Brokensha, D. and J. Glazier. 1973. Land reform among the Mbeere of central Kenya. *Africa* 43(3): 182-206.

Brokensha, D., B.W. Riley and P. Castro. 1983. *Fuelwood Use in Rural Kenya: Impacts of Deforestation.* Institute of Development Anthropology. Binghamton, New York.

Agroforestry Potentials and Land Tenure Issues in Western Kenya

Noel A. Chavangi

Kenya Woodfuel Development Programme

Introduction

The information on issues discussed in this position paper has been collected through field investigations in Kakamega District on existing agroforestry practices, the utility and consumption of available woody biomass, and the socioeconomic forces within the household which mediate domestic production and the utilization of woodfuel. This is intended to form the basis for the design and subsequent testing of technical options and extension strategies for the development of effective woodfuel production projects in the District.

The need for increased woodfuel production was perceived because of energy policy planning activities in Kenya that started in the late seventies and whose major events include the Kenya National Energy Symposium (November 1978), and the International Workshop on future energy policy issues for Kenya (May 1979). After the creation of the Ministry of Energy and Regional Development in 1979, the Ministry invited the Beijer Institute of the Royal Swedish Academy of Sciences to assist in carrying out a systems analysis that would form the basis for programs on energy provision in the future.

The analysis carried out is now known as the Kenya Fuelwood Cycle Project whose conclusions have been adopted by the Ministry which has since embarked upon a number of studies and projects with a view to translating the broad conclusions of the study and the indicated options into intervention programs. The Kenya Woodfuel Development Programme (KWDP) is one of the projects operating under the auspices of the Ministry. The overall objective of KWDP arises from one of the main findings of the above study, to the effect that woodfuel (fuelwood and charcoal), as both a non-commercial and commercial source of energy, accounts for about three-quarters of the total energy base of Kenya and that the scope for employing other locally-based energy sources is limited. In addition the economic trend is unlikely to encourage a move from woodfuel to other fuels (like oil and coal). As a consequence, woodfuel will continue to play a dominant role within the energy economy of Kenya.

Fuelwood accounts for 53 percent of the total energy consumption in Kenya and is dominant in the rural areas (72 percent of the total fuelwood consumption in Kenya). Charcoal accounts for 6 percent of the total energy consumption in Kenya and is dominant in the urban centers (50 percent of the total charcoal consumption in Kenya). Fuelwood is mainly produced and consumed in the high and medium potential agricultural areas, while charcoal is produced in the semi-arid areas and consumed in the urban centers.

The greatest proportion of Kenyans live in the high and medium potential agricultural lands which have a high degree of land demarcation and privatization, offering almost no possibilities for the promotion of communal fuelwood plantations. Efforts towards the supply of energy for the millions of Kenyans living in these areas have to concentrate on the promotion of increased tree planting activities on individual household lands.

Issues relevant to land tenure and trees

Information gathered on socioeconomic and cultural aspects of land tenure and treeplanting activities have raised several prominent issues; these are considered most relevant for discussion at this workshop.

Tree planting activities in relation to land ownership

For practical purposes, land ownership in Kakamega can be said to exist on a private tenure basis supported by both legal and traditional provisions and basically falling under ancestral land transactions with options of land sales. Generally the right of allocation of use is entrusted to the recognized male head of the household who will usually make consultations with other members of the family before taking final decisions.

Private rights to land ownership are therefore very strong. Trees are considered as permanent features of the land (possibly a reflection of the long time trees take to mature). Traditionally, it is only the owner of the land who can plant trees, since tree planting may establish a claim to land ownership through tree ownership. In Kakamega, land disputes have traditionally been resolved in favor of the party claiming ancestral ownership of the most mature trees on the piece of land and especially so where such trees may be growing on a site of an ancestral grave, or a place with cultural value in relation to certain ritual requirements.

Women in relation to tree planting activities

According to tradition, women in Kakamega do not have any permanent rights to land ownership and therefore, since trees are considered in a permanent relationship to land, tree planting has been traditionally considered as a "taboo" for women. This has been used over the years to rule out the possibility of women claiming ownership of land through tree planting activities. The stength of this taboo varies between households according to such factors as religion, literacy, etc. Although the taboo has weakened for some women, its current strength and prevalence should not be underestimated.

Implications for fuelwood procurement

According to tradition, fuelwood gathering is basically a women's task. Men rarely associate themselves with this activity in Kakamega. Fuelwood gathering has for a long time been conceived as the procurement of a common and free good, taking place on communal or private lands as each different local situation allowed. Access to communal or private lands for fuelwood is diminishing as land becomes scarce because of population pressure. There has been, however, no appreciable movement by farmers toward deliberate tree planting activities specifically for fuelwood.

The women, therefore, depend for their fuelwood needs mostly on the household lands to which they have access, e.g. hedges, remnants of bush, etc. In the high potential smallholder agricultural areas where all the available farm land is under cultivation, acute fuelwood shortages have resulted in the commoditization of fuelwood. The situation is aggravated by the fact that women, by tradition, do not participate in tree planting activities and therefore have no direct rights to trees and tree products. The production of fuelwood on the farm is therefore seen as the *missing link* in the fuelwood procurement cycle.

Possibilities of tree planting for fuelwood

Despite the nature of the prevailing situation as outlined above, self-sufficiency in fuelwood supply may be achieved on a sustainable basis if women can be given the opportunity to participate in fuelwood production activities. Indeed a case in point exists in parts of Kakamega (Bunyore) where *Sesbania sesban* is planted by women and is generally referred to by both men and women as a fuelwood species.

In the most densely populated parts of Kakamega District, some areas are supporting more than 1,000 people per sq. km, with extremely high rural homestead densities (185 per sq. km) and correspondingly small farms, averaging 0.53 ha (among the most densely populated areas in Kenya). Land is used very intensively with an overwhelming concentration of directly consumable rather than commercial crops (e.g. maize, beans and bananas).

Despite the intensive pressure on land, a significant proportion is devoted to the production of woody biomass (21.2 percent), represented by all the possible tree and shrub configurations (wood lots, hedges, windbreaks, individual trees, etc.). Planted trees and shrubs account for up to 55 percent of all woody biomass.

Paradoxically, the high percentage of woody biomass is not reflected in adequate supplies of fuelwood at the household level. The trees are categorised as men's cash crops and therefore make a minimal contribution to the household fuelwood needs. There is in fact a high degree of wood commoditization.

Observations indicate that if an approach to fuelwood supply were to focus on the diversion of the present planted woody biomass from the market to meeting the household fuelwood needs (a very unwelcome proposition to the men), the ultimate result would be a decrease in the supply of tree products on the market, leading to an increase in prices that may further aggravate the fuelwood shortage situation at the household level for those who purchase fuelwood.

It is therefore necessary to consider the possibility of planting quick-growing fuelwood tree species in addition to, or in replacement of, part of the existing woody biomass. It is estimated that 1,000 fuelwood trees are needed annually to make an average family of eight self-sufficient in fuelwood. The question is, how can this be achieved given the small farm sizes prevalent in these areas (average sizes of 0.53 ha). KWDP has identified three major action areas requiring attention at the household level:

- the development of an awareness program through which farmers can be motivated to assess the need to plant fuelwood trees;

- the development of viable field tested agroforestry technical options that can be encouraged for fuelwood production;

- the development of extension methods and strategies that can effectively relate to farmers in a manner that leads to the strengthening of the farmers' participation in solving the fuelwood shortage problem.

The basic consideration is the provision of a system through which farmers are given an opportunity to indicate their own perception of the situation and deliberate on possible solutions as seen by them.

KWDP is approaching the issue through the involvement of the local community in efforts towards working out possible local solutions. Group discussions (of about ten to fifteen villagers, both men and women) are held to consider problem-solving approaches, and results so far are quite promising.

Issues arising out of the group discussions

Definition of a tree

It appears more and more certain that based on the "frame of reference" of the group discussion participants the definition of a "tree" in relation to land and utility may be a significant factor. It is becoming apparent that certain short term maturing tree species that are of major importance in agroforestry (e.g. *Sesbania sesban*, *Lantana camara*, *Tithonia sp.*, etc.) *may not be considered "trees"* by certain segments of the Kakamega population. These species are seen more in the context of "fuelwood shrubs," "fodder for cattle," etc.

Trees and land demarcation

Some other tree species play an important tenure-related role as legally recognized land demarcation species in addition to their function as a hedge. *Euphorbia tiracali* is one of the most important species for land demarcation purposes, but has the disadvantage of not contributing much to the household's fuelwood supply. It appears that in some areas euphorbia has been replaced by other species (*tithonia*, *lantana*, etc.) which play a more important role in the supply of fuelwood to households.

All these observations indicate a possible breakthrough in the promotion of fuelwood species outside the "permanent tree" concept. KWDP is currently working on this possibility using known fuelwood tree species (*Sesbania sesban*, *Mimosa scabrella*, *Calliandra calothyrsus*, *Leucaena leucocephala*, etc.). The added advantage of possible improvements in soil fertility by these species has been found to be worth considering. The implied "temporary land occupancy" aspect of these short rotation species may also facilitate the direct participation of women in the planting of fuelwood in various agroforestry configurations as part of their existing farming activities.

Local suspicion of tree planting programs

Farmers are very suspicious of any tree planting programs. There is a feeling that the promotion of tree planting activities by agencies may be a maneuver to claim the land from them or deprive them of the right to cultivate the land on which the trees are planted (this may also be applicable to other crops).

Implied participation by women

In narrowing down to the current focus on selected fuelwood species, emphasis has been placed on the role of women in fuelwood supply to the implied exclusion of male participation. As a matter that concerns the allocation of resources and responsibilities at the household level this is an area of primary conflict. The foregoing exceptions notwithstanding, tree planting is still largely defined as a male activity. How can women's participation be ensured? One factor which may make it easier for women to take a more active role in farm management, including fuelwood planting, is male migration to the towns in search of employment, especially in the densely settled areas where farm sizes are simply too small to support an average family of eight.

Conclusion

KWDP activities focus mainly on the possibilities of increased tree planting for fuelwood supply. The farmer, however, approaches the issue of tree planting on a wider basis, giving more emphasis to the multipurpose aspects of trees, and with a major inclination toward trees for construction purposes and sale. Agroforestry projects have to accommodate most of the needs of the farmer, as indicated by the community in general and the individual households in particular. However, KWDP may have to focus on the participation of the women, inasmuch as women's responsibility for fuelwood supply at the household level is still very significant.

The challenge is to develop fuelwood production systems that take into consideration all the issues of land tenure, tree ownership and women's participation as highlighted in this paper.

References

Bradley, P.N. 1984. The district resource analysis as applied to Kakamega District. KWDP Working Paper No. 2. KWDP. Nairobi.

Bradley, P.N. and N. Chavangi. *In press.* Development research and rural energy planning in Kenya: Part II. From the district to the farm. *Ambio* 14.

Bradley, P.N. and A. van Gelder. *In press.* Development research and rural energy planning in Kenya: Part I. The district focus. *Ambio* 14.

Chavangi, N. 1984. Cultural aspects of fuelwood procurement in Kakamega District. KWDP Working Paper No. 4. KWDP. Nairobi.

Engelhard, R. 1984. The origins and development of the Kenya Woodfuel Development Programme. KWDP Working Paper No. 1. KWDP. Nairobi.

Gelder, A. van and P.N. Bradley. 1984. The search for viable development development options. KWDP Working Paper No. 5. KWDP. Nairobi.

Gelder, A. van and P. Kerkhof. 1984. The agroforestry in Kakamega District. KWDP Working Paper No. 3. KWDP. Nairobi.

Gelder, A. van and P. Kerkhof. 1984. Agroforestry survey results. KWDP Working Paper No. 6. KWDP. Nairobi.

East African Pastoral Land Tenure: Some Reflections from Maasailand

Barbara E. Grandin

International Livestock Centre for Africa, Kenya[1]

Pastoral production and land tenure reform

Non-veterinary rangeland development programs in East Africa have generally focused on stemming degradation and on increasing the commercial offtake of pastoral production. Land tenure reform is often mooted as an essential ingredient of proper rangeland management and hence, pastoral development. Land tenure in East African pastoral areas was traditionally communal and to a large extent it so remains. Rights to grazing are obtained by virtue of membership in a social unit. Due to ecological variability and erratic rainfall patterns, traditional access areas and their concommitant social units are large. The social unit sharing an area (the access group) may be a clan as among the Samburu, a section/sub-tribe as among the Maasai, or some other locational grouping. Below the access group there are smaller groups controlling local resources. Figure 1 presents a schematic of socio-spatial organization in East African pastoral areas, outlining the segmentary structure (from the ethnic group to the household) and the primary functions of each level.

The urge to reform land tenure is based largely on an acceptance of the concept of the "tragedy of the commons." This, in turn, is based on the assumption of human greed (manifested in unnecessary herd accumulation), which is unfettered by any control of access to the basic resources of grazing and water. This concept, and therefore much of range development, fails to appreciate several essential characteristics of pastoral production.

First, most pastoralists have insufficient livestock to maintain the desired quality of life. The marginal status of pastoralists (spatial, economically, and physically) necessitates the keeping of large numbers of animals for short- and long-term self-sufficiency in the face of periodic droughts and epidemics. Despite ideologies of equality, pastoral holdings vary tremendously in size (Grandin 1983), frequently by a factor of 10 on a per capita basis. Second, traditional social mechanisms existed for the distribution of products in pastoral areas.

SMALLEST **A.** Household production unit

-locus of cattle ownership
-autonomous decision-making
-highly mobile
-flexible; may split seasonally
-viability problem (labor/animal balance)

B. Joint residential unit (compound/homestead)

-joint unit for herding/watering
-strong prescription for food sharing
-domestic self-help unit

C. Local unit (e.g. neighborhood)

-broader cooperation/information exchange, sociability
-share/control local grazing and water resources
-often core nucleus population with regular influx/outflow of others

D. Primary resources sharing units (access group)

-large to allow for resource fluctuations
-theoretically free access to all members
-largest unit of traditional administration

LARGEST **E.** Political society/ethnic group

-ideological unit
-shared language and culture
-limited access throughout areas in times of severe stress
-occasional hostilities

Figure 1. East African pastoral socio-spatial organization (a schematic).

Those who had surplus animals were obligated to care for those without sufficient livestock (i.e. food flowed from rich to poor while labor flowed from poor to rich). Strong pressures for food sharing which typified pastoral culture are being eroded by the pastoralists' incorporation into the market economy and by the move towards individuation in both production and consumption (fostered by programs aimed at increasing commercialization). Factors of production, especially animals, and now increasingly labor, are traded through market channels. If individuated, land will follow suit, and landlessness will undoubtedly occur.

In regard to uncontrolled access to resources, as assumed by the "tragedy" argument, there is evidence that social mechanisms to control access to grazing and other resources have existed at various levels of socio-political organization. Despite claims to the contrary (Strange 1980), communal tenure does not necessarily imply "uncoordinated use". However, in many areas, the traditional socio-political organization which formed the basis of control has been perceived as a threat to national governments and has been systematically undermined. Despite the erosion of traditional systems, social control of access to some resources continues to be important.

Rangeland planning is often dominated by natural scientists who have been unaware of traditional control mechanisms or unsure of how to strengthen them. Individuation of tenure is often seen as a panacea; perhaps the first and most essential step in stemming the tide of range degradation, by making people feel "responsible" for their own resources by confining individuals to their own piece of land. This philosophy, of course, ignores the widespread occurrence of resource deterioration under individual tenure in other systems.

The group ranch experience

Group ranches in Kenya are frequently mentioned as a successful land tenure intervention (although they are not deemed successful in commercializing production). Through group ranches, communal lands were divided into smaller units (ranches) with officially registered members holding a group title deed and intended to make exclusive use of the ranch resources. Group ranches were viewed as a compromise between individual ownership (which characterized much of Kenya's agricultural sector) and the need for access to wider resources in dry lands.

The Kaputiei section (sub-tribe) in Eastern Kajiado district, was the first group ranch to be adjudicated. Traditionally the entire section was a simple access group, controlling an area of 3100km ^{2}, although it had two subsections (north and south) of 960km ^{2} and 2140km^2, respectively (see Table 1). A number of elite Maasai were given large individual ranches of 300-800 hectares. In some northern areas, producers refused group ranches and were given individual tenure to holdings of usually less than 100 hectares. The rest of the north and the bulk of the south was divided into 15 group ranches, averaging only 163km^2. It is now clear that: 1) group ranches did not coincide with any level of traditional socio-political organization, 2) the new mode of organization (i.e. group ranch committees) has been largely unsuccessful, 3) confusion regarding the locus of authority and lack of coordination have ensued, and 4) movement across ranch boundaries remains a common response to dry conditions (Grandin 1981).

Table 1. Kaputiei Section, Kajiado District

	North	South	Total
Size (km^2)	960	2140	3100
Group ranches (no.)	3	12	15
Avg. size (km^2)	151	166	163
Avg. membership1/	143	120	126
Individual ranches (no.)2/	690	13	703
>3 km^2	75	11	86

1/ As of 1980
2/ Source: A.H. Jacobs personal communication 1984

In the better watered areas of Maasailand (the majority), individuated tenure is becoming the norm. Many areas which initially accepted group ranches are pressing for subdivision. Other areas, developed at a later phase, have opted to move directly to individual tenure. The desire for individual tenure stems from a variety of factors including: 1) wanting a title deed as collateral for loans (which are denied group ranchers as individuals), 2) frustration with the inefficiency of the organization and management of group ranches, and 3) fear of eventual land loss (enhanced by the government's inability to control squatting on group ranches). In addition, the haste for subdivision is fueled by a burgeoning population; the more adult men there are, the more registered members there will be, and the smaller the individual's share of the land. While individual tenure is ecologically feasible (if not optimal) in much of Maasailand, it would be disastrous in most other, drier, pastoral areas.

The pressure for individuation of tenure raises major problems in East African pastoral areas for several reasons. Most of these areas are characterized by relatively low potential and erratic rainfall (marked by extremely irregular distribution across space and time). In addition to rainfall, other production resources (natural salt, wet and dry season water, etc.) are scattered through space. Thus the larger the land of access, the more efficiently it can be exploited in the long run (by periodic mobility rather than destocking or death).

Uncoordinated use: a myth about pastoral resource management

By assuming that communal land tenure entails uncoordinated use, planners have frequently neglected to investigate the differential tenures and use of grazing and residential areas in pastoral lands. Studies by ILCA in Maasailand have demonstrated that land was delineated into residential areas, wet season grazing areas, early dry season and late dry season grazing areas (Peacock et al. 1982). Neighborhood groups controlled movements in their area. In addition, there were drought "fall-back" areas (frequently far from water or disease infested). Because these fall-back areas were not normally exploited, they were defined by the wider polity as "unused" and were frequently alienated. Within the residential areas, groups of households set aside areas of land near the residential sites which provided reserved grazing for calves too young to walk long distances, and for sick or weak animals. Healthy adult animals were not allowed access to these special pastures. Today, in Kenya's Maasailand these calf pastures are becoming individually held and are forming the basis for land claims in the press for subdivision of group ranches.

A. Traditional Subsections Before Group Ranches

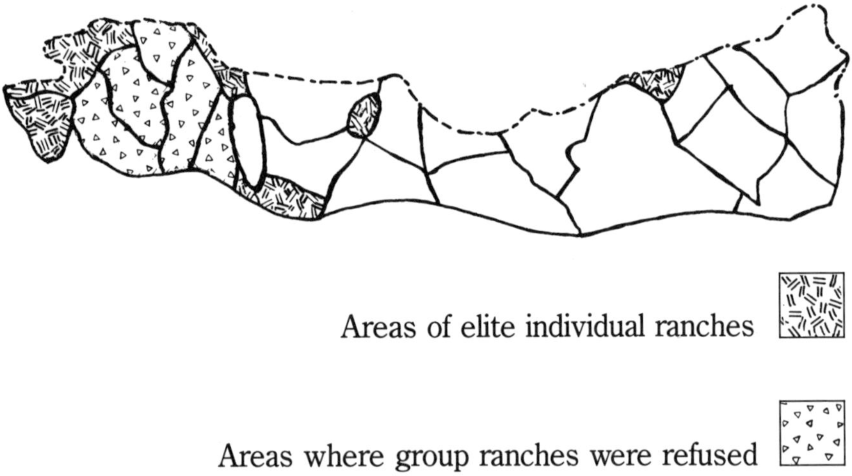

B. Group and Individual Ranches

Figure 2. Territorial organization of Kaptitei section.

In pastoral areas, as in many agricultural societies, inputs of labor give a production unit special rights to resources. In Maasailand, the digging of a well gives the "owner" special control over the water, and to a certain extent the well area. Well locations were "inherited" if they did not fall into disuse. Today, as Maasai are gradually moving into crop production, the fencing and plowing of a field gives the production unit temporary control over this piece of land.

Trees have traditionally not been planted in Maasailand; they are considered a common property resource. Trees on calf pastures have some special rights attached to them. Permission is not required to cut dead trees for firewood, or to cut a few living branches (e.g. for a walking stick), or to take a few roots for use as herbs. The only time the permission of the controller(s) of the calf pasture must be sought is when cutting massive amounts of branches for fencing. Improved housing, waterpoint development, establishment of fields, and to some extent, tree planting, are used in Maasailand to stake land claims in anticipation of subdivision.

The need for community based development in pastoral areas

Given the relatively low population densities in pastoral areas provision of services (veterinary, credit, etc.) is difficult at a household level. This difficulty is compounded by mobility patterns. Infrastructures (e.g. water points, dips) will not be economic for a household to use alone. Some form of cooperative organization for access to resources and services would appear to be essential. These should be based as much as possible on existing institutions and cultural patterns of decision-making and sharing of responsibilities. Equity concerns, however, should not be ignored.

Understanding local culture and social organization will provide important guidelines for fostering community responsibility and action. It will be useful to identify groups of people who have sufficiently long-term mutual obligations that could form the building blocks of cooperative effort. Social organization is marked by spatial ties (compound, neighborhood, community) and by cross-cutting social ties (affinity, clanship, age-set). Many of the strongest ties in pastoral systems are the social ties that facilitate mobility, but often at the expense of strong locational ties at a low level. Creativity will be required in order to use rather than ignore these relationships for pastoral development.

Many proposals for "proper" (i.e. Western) range management have assumed that a significant portion of the ever-increasing pastoral populations would have to leave the land to find alternative means of livelihood. However, all indications are that this will not always be possible. In Kenya, there has been significant inmigration to better watered pastoral districts. Between 1969-1979, while the nations total population grew by 59 percent, the two best watered districts under pastoral control (Narok and Kajiado) had a growth of 74 percent (Jacobs 1983). Plans for development in pastoral areas must accept the challenge of increased population, which means more people to feed and more labor available. Extensive, western style ranching systems are largely inappropriate (Grandin 1985). As Boserup and others have shown, increased population pressure often leads to an intensification of agricultural production, mainly through increased labor inputs. Development pathways should be sought for pastoral areas to utilize this labor. Agroforestry, with an emphasis on dry season fodder production, should be explored as one source of such pathways.

Notes

1. The views expressed in this paper are those of the author and do not necessarily represent those of ILCA.

References

Grandin, B.E. 1981. Group ranches in Kaputiei: The impact of the Kenya Livestock Development Project, Phase 1. Internal Report. ILCA. Nairobi.

Grandin, B.E. 1983. The importance of wealth effects on pastoral production: a rapid method for wealth ranking. In *Pastoral Systems Research in Sub-Saharan Africa*. ILCA. Addis Ababa.

Grandin, B.E. 1985. Human demography and culture--factors in range management. Paper presented at the conference on Rangeland Resources: Wildlife/Livestock Interfaces. Taita Hills, Kenya.

Jacobs, A.H. 1983. An overview of population dynamics in Kenya's rangelands (1969-1979). Winrock International. Morilton, Arkansas.

Peacock, C.P., P.N. de Leeuw and J.M. King. 1982. Herd movement in the Mbirikani area. Internal Report. ILCA. Nairobi.

Strange, L.R.N. 1980. Human influences in African pastureland environments. FAO Pasture and Fodder Crops Studies No. 8. FAO. Rome.

Land and Tree Tenure Issues in Three Francophone Sahelian Countries: Niger, Mali and Burkina Faso

James T. Thomson

Associates in Rural Development, Inc., Burlington, Vermont

Main tenure problems and their impact on adoption of agroforestry innovations in Niger, Mali and Burkina Faso

Two tenure issues hamper agroforestry at the moment: forestry code provisions determining tree tenure rights, and national government assertions of authority over the land base in the three subject countries. Of the two, the first is immeasurably more serious in most settings. Foresters do actively intervene in the establishment, enforcement and manipulation of tree tenure rights. By contrast, governments expropriate lands belonging to individual peasants far less often, even though expropriation typically figures as a major problem at the edges of expanding cities and in some project areas, particularly those involving dams, irrigation perimeters and creation of new state forests.

Forestry code provisions in each country define some ten to fifteen major tree species, wherever sited, as common pool resources, nationally regulated under provisions of a common property regime. Details vary from country to country, but the overall effect is the same: to the extent that the forest service is capable of making its presence felt, local-level tree tenure rules are placed in question or superseded entirely by State regulations.

In theory, to order to trim, lop or cut for personal use a live tree of a protected species (on the cutter's private land, on common land, in state forests, etc.), the individual must first get from a forester a free authorization which renders his action legal. Cutting for commercial purposes is governed by roughly the same system, except that fees are charged for the different types of permits.

In practice, management of the woodstock under a unified, national-level common property regime breaks down badly. The few exceptions, where controls are adequate, typically involve state forests with rich stands of trees that, from a forest service perspective, merit protection. *Outside state forests, however, the norm is inad-*

equate control. Because foresters have claimed tenure authority over the best tree species, locals do not often organize on a long-term basis to control access or usage. Rural dwellers do not provide patrol and surveillance services in the area of woodstock protection. Instead, everybody harvests on a need basis. Most users take what they need or think they can get away with. People adopt strategies calculated to reduce personal risk, i.e., they harvest, not on their own lands, but on remaining common waste lands or on the property of others. In effect, the formal, nationally-established common property regime generates a set of incentives for individual users which leads them to convert a common property resource into an open access resource at the local level.

Incentives to invest are largely non-existent. In areas where the woodstock has long been abundant, but is now coming under pressure, the assertion of national authority over the most valuable trees hinders the development, often observed elsewhere, of private property or more vigorous local common property regimes, which could generate incentives promoting investment. Instead, where the woodstock is already scarce, people frequently hesitate to plant trees or -- far more relevant in the arid zones of these countries -- to protect natural regeneration. In consequence, efforts to promote agroforestry appear hampered because people seriously question their ability to protect and exploit as suits them the trees they produce.

Agroforestry systems showing potential for solving or mitigating existing tenure problems

Central Sahelian agroforestry systems can be divided into two rough categories: those which are based on local tree species, and those which rely on exotics. The former typically involve valuable species protected under forestry code provisions, e.g., *Acacia albida.* Such trees -- where scarcity is perceived -- would usually be protected by local people within the limits of their means. At the moment protection is lax because villagers cannot with any ease directly influence the evolution of tree tenure rules toward forms they would consider more suitable.

Agroforestry systems based on exotics, principally neem (*Azadirachta indica*) and imported fruit trees, e.g., mangoes, etc., tend to provide villagers more liberty from state control because the trees clearly do not belong to the class of protected species. Foresters by and large accept the trees that individuals plant in compounds, gardens and fields, as the private property of the planters. Unfortunately, the surface area involved is limited, particularly that in fields.

Mixed systems have begun to evolve, often at donor, PVO and/or forestry service initiative. Border plantings and windbreaks, e.g., the CARE/Eaux et Forets windbreak project in Niger's Majjia Valley, have been designed with an eye to multiple-use development. Successful management of the mix of exotic and local trees may eventually convince decision makers that rural residents will invest in reproduction of the woodstock if the right property incentives are provided.

Promising new approaches to the social organization of land use and their relevance for agroforestry.

Here again two developments stand out. The first merits particular attention. It involves attempts by local individuals, and sometimes by whole communities, to create "outlaw" tree tenure systems at the local level without formal state sanction. At various sites in Niger, Mali and Burkina Faso, local people are asserting or reasserting private or common property rights to trees. Often these involve actions to protect live trees in fields as private property, or to recover wood harvested by the field owner and appropriated by another without the harvester's permission. Sometimes these local initiatives appear to involve groups, who decide to manage the woodstock on village lands as an integrated unit.

All of these efforts indicate a willingness of local people to bear at least some of the costs of managing the woodstock in their area. Particularly in those situations where an entire local community or group of communities has decided to control exploitation of the woodstock, the potential for development of a more broadly-based set of land use policies is apparent. When and if these local perceptions are coupled with concerns about controlling movements of foraging animals during the dry season and watershed management, it should be possible to move forward to more active efforts to preserve and enhance environmental potential under both common and private land tenure.

In several instances, forestry projects have taken the initiative in involving wood users in woodstock management. In Niger, the USAID Forestry and Land Use Management Project has moved to put a degraded state forest near the national capital on a sustained yield footing, in part through regulating browsing and access to arable lands within the forest in return for compliance with prescribed procedures for woodstock enrichment, e.g. windbreaks, shelterbelts, infield plantings of *A. albida*, etc. This system may eventually be extended to other areas in Niger. In Southern Mali, near Sikasso, the Swiss-financed forestry project appears on the verge of shifting from a strategy of promoting village woodlots to one of promoting protection and sustained-yield management of natural vegetation on small plots of village land. These and similar developments are new; it is difficult at present to estimate their potential for success.

Impact of new tenure and agroforestry developments on disadvantaged groups

In the contemporary Sahel, landlessness still appears to be a minor problem, though probabilities are high that it will rapidly become more serious. Although in some localities, shortages may be serious in regional terms most individuals who want access to "productive" land can get it. The pertinent problem is, rather, the poor quality of the land, often exacerbated by the unpredictable rainfall regimes characteristic of arid areas. In this sense, probably upwards of half of current Sahelian populations qualify as "disadvantaged." Women may well suffer more than men in this regard.

Over the long haul, and assuming population can be brought into balance with the sustained-yield potential of renewable Sahelian resources, agroforestry practices are indispensable. Their connection with both fertility of specific production sites and generalized economic management of the fragile ecologies characteristic of the area is patent.

Research needed to clarify particular tenure issues in agroforestry

Applied research to demonstrate the feasibility of local, or joint local/forest service management of woodstocks is imperative as a step toward involving rural Sahelians in active efforts to preserve, manage and exploit natural vegetation on a sustained-yield basis. In essence, applied research must demonstrate that rural people can be trusted to use trees without destroying them. Once that has been empirically established, it will be possible to consider revising forestry codes and working out more sophisticated approaches to agroforestry which will upgrade the resiliency of the environment in the face of drought and growing population pressures.

One of the fundamental issues to be resolved concerns the types of tree tenure systems most appropriate in particular settings. In some areas for instance, the state-created, nation-wide common property tree tenure regime should probably be maintained, e.g. in state forests, which often double as game reserves.

In many other situations, the formal common property regime should almost certainly be privatized to encourage better management, i.e. disaggregated into small parcels owned and controlled by individuals or small family groups. Such a scheme might work rather nicely in those settings where dispersed dwelling patterns make it relatively easy for those who own fields to keep them, and the woodstock on them, under surveillance.

In all likelihood this approach would unify land and tree tenure rules, so that field owners would also own all trees occurring naturally on their lands. This strategy involves the risk that landless individuals, or those whose lands currently contain very few trees, may suffer severe costs if forced to purchase their wood supplies from those who own a larger supply. Relationships between land owners and non-owning land users on this point would also have to be worked out, presumably when agreements are negotiated, or as a result of dispute resolution processes in local tenure systems. Nationally-imposed solutions, though perhaps designed to protect those less well off, might well prove cumbersome and ill-adapted to many local situations.

In other areas, the state-organized commons should probably be disaggregated into a series of locally organized and managed commons. Officials of these local organizations should be authorized by appropriate legislation to make and enforce rules governing access to, and use of, the common property woodstock resources. This approach could preserve the access and use rights of those local residents who now harvest wood and minor forest products in the area, even though they may have few or no trees on their own lands. At the same time, it poses the often thorny problem of organizing effective control and enforcement mechanisms. In many Sahelian villages, where labor migration has largely emptied communities of their youth, it may be extremely difficult to recruit enough individuals to mount surveillance patrols to police access to and use of the woodstock.

These issues are further complicated by restrictions which presently bear on rural government activities in Niger and Mali (the situation in Burkina Faso apparently remains in flux because of initiatives adopted by the new regime). In both those countries, local groups cannot freely organize, nor can such groups now obtain formal state authorization to levy taxes to support the cost of such local governmental activities.

Whatever the policy decisions on these issues, and the activities subsequently initiated to test proposed solutions, very careful monitoring should be undertaken to determine the consequences of different tenure systems on the quality of woodstock management and environmental protection.

Relationship of the above considerations to planning and implementation of research and development projects

Both direct research activities, and projects through applied research sections, should focus consistent attention over the next decade on promoting and monitoring local management efforts. These should include a variety of property regimes: some state, more private and common, and perhaps some open access regimes in areas of remaining abundance.

Efforts should be made to incorporate local legislative capacity into the design of forestry and agroforestry projects. Resource users should be given every encouragement to experiment with new approaches to regulating exploitation of and investment in the renewable resources--soils and trees--that lie at the heart of agroforestry systems. Such activity would be appropriate in both common and private property regimes, and might become appropriate in future for what are now legally or *de facto* open access regimes.

To encourage such activities, projects might well consider offering local resource users a series of options regarding the kinds of organizational structure and legislative regimes they might adopt in order to acquire greater control over their local resources. Research should closely examine the patterns of choice in different areas, both to determine weaknesses of various options and to identify those options which most effectively promote beneficial agroforestry developments.

Gum Hashab and Land Tenure in Western Sudan

Abdul Gasim Seif el Din

Khartoum, Sudan

After long years of unrest and tribal wars, which partly resulted from the immigration of Arabs into Sudan, the peoples of this country finally settled into the tribal groups of the present day during the period 1400-1500 A.D. The land of each group or tribe was recognized by others through land marks and members of the tribe enjoyed the rights of cultivation and grazing of their animals within the specified boundaries. The nomadic tribes, on the other hand, were allowed by agreements among tribal notables to move through lands of other tribes, thus enjoying privileges over a much larger area than prescribed to their own tribes.

When the gum trade began to flourish in the latter part of the 19th Century (Hill 1959), farmers in Kordofan began to claim ownership to gum forests. Each man claimed as much land as he could lay his hands on by cutting down some trees to demarcate the boundaries of his claim. As might be expected, this rush for land caused disputes which often led to bloodshed. To settle these disputes, the tribal leaders laid down rules to organize land tenure in their domains and this resulted in the rise of local customs and rules on land titles which varied from one tribe to another. These rules gave members of the group the right to own a piece of land for cultivation and gum collection and also the privilege of gum collection from the unallocated communal lands.

The first trial of government intervention in organizing land tenure was made by Kitchener who issued the Title to Lands Ordinance of 1899. With this he aimed at settlement of the people to increase agricultural production, which had been badly disrupted during the wars. The ordinance recognized the validity of titles that were acquired during the Mahdia (Warburg 1971) and ruled that continuous cultivation during the five years preceding the date of the claim created an absolute title to the land.

The gum tree, *Acacia senegal* (hashab), regenerates freely on cultivated lands, and when a piece of land is left fallow after cultivation a gum garden is formed (Jackson and Shawki 1950, Seif el Din 1969).

Large numbers of such gum gardens sprang up as a result of shifting cultivation, leading to the present bush fallow system. With increasing importance of the gum trade, the peasants put more effort on land acquisition and, whenever circumstance permitted, new additions of land were made from unallocated areas. Those who owned baobab trees (*Adansonia digitata*) had better chances of acquiring gum gardens because the water stored in these trees was an important asset during the dry season.

In central Kordofan some land owners made use of landless strangers to establish new gum gardens. Those strangers (dandara) usually came from tribes which had settled in the more arid northern areas. They were allowed to cultivate a piece of land for a period of three to six years, i.e. until it was exhausted, after which it was handed over to the owner, usually containing a growing thicket of hashab. This process was repeated on other plots until all the land was regenerated with hashab.

Here no acknowledgement was made of the Land Ordinance of 1899 which specified that a continuous cultivation for five years gave the cultivator undisputed title over that land. This was mainly because villagers among themselves recognized land rights as being established by merely making a mark on any tree to demarcate a boundary. This form of land acquisition resulted in some people having large holdings of gum gardens and others having very little or none. Those who possessed more gum gardens than they could tap depended on hired labor for gum production. Some owners let out parts of their holdings on a half-share basis or for an annual rent known as "tugundi," a term which was later used by the government for the tax levied on gum gardens.

Since the beginning of the present century, several schemes have been proposed to settle disputes concerning land and gum gardens. Tippetts (1908) proposed that all gum gardens in eastern Kordofan be registered after the claimants proved their rights and that each owner should clear a fire line 5 metres wide around his garden. The lands over which no rights were claimed were to be declared as government property, and rights of gum collection to be granted to applicants upon payment of a small annual rent. The registered gardens and the government lands, Tippetts suggested, should be published in the Sudan Government Gazette. Craig in 1912 thought that Tippett's proposal would consume too much energy and, instead, suggested that only boundaries of village lands be demarcated and the rights of cultivators over their fallow "bura" be recognized.

Nicholls (1913), after making a study of the problems pertaining to rights of gum collection, proposed that the following procedure be adopted:

1. To be registered to government:

   a) all rights over "ghifar" (unallocated land);
   b) all rights over gardens lying outside the registered village boundaries;
   c) all rights which are not included in (2) below.

2. To be registered to the individual inhabitants of villages:

   a) rights over all gardens in their own village which they can prove were originally "bilad" (cultivated by themselves or their immediate ancestors);
   b) rights over all non-ghifar gardens in their own village to which they have established a sound prescriptive right by absolute possession since the resettlement.

Nicholls also proposed that transfer of gardens by sale or gift should not be recognized except between inhabitants of the same village in respect of gardens situated in that village. Government gum gardens were to be under the management of tribal heads, acting as govenment agents and receiving 10 percent of the annual rent of those gardens. When anyone left the village to settle elsewhere, his garden was to be at the disposal of the government, to be given to another villager, a new settler, or registered to the government.

Willis (1912) made proposals similar to those of Tippetts as regards land ownership in western Kordofan and added that the collection of "tugundi" was to be discouraged. He further added that people originating in other villages who tap gum gardens for tugundi ("dandara") should be considered members of the village and the gum gardens should be registered in their names. Baobab trees, as already mentioned, were an important element in the growth of settlements and utilization of gum gardens in western Kordofan, because these trees were hollow and used to store water for the dry season. Willis, therefore, proposed that demarcation of village lands should include settlement of disputes over baobab trees lying between any two villages.

Various officials of the government made proposals similar to those outlined above in an attempt to settle disputes over gum gardens. Land and gum problems, however, remained unsettled and disputes which started in that time are still heard in courts today. Newbold (1935), then Governor of Kordofan, described the system as "an elastic but ordered system of communal tenure" and proposed that "individual registered ownership with its attendant social and economic evils in tribal communities" should be avoided. In order to ensure continuation of the communal tenure, Newbold suggested that the native authorities should have a uniform and effective policy of land administration.

The Land Settlement and Registration Ordinance of 1925 stipulated that all lands which were not registered in the names of normal persons were regarded as government property but that rights of cultivation, pasture, forest produce, occupation and any other benefits by individuals were recognized over such lands (El Mahdi 1971). The right of occupation, which meant the process of opening up part of the natural forest for cultivation, was particularly favorable to those who were able to lay their hands on gum lands that were otherwise unoccupied.

This ordinance, however, was not effective as a system of land tenure among the rural populations which responded more readily to the customs and tribal rules. This is only natural because the tribal leaders preferred to deal with their own peoples in their traditional ways in preference to new laws that were not only difficult to implement but also restricted their influence on their tribes.

The Unregistered Land Act of 1970 reaffirmed the government's ownership of all unregistered lands by declaring that such lands be regarded as fully registered in the name of the government. This means that all gum gardens are registered to the government, but the Act recognizes the rights and privileges enumerated in the LSRO of 1925. This series of laws was an attempt to find a way of putting land tenure in some legal context but, due to the complexity of the Sudanese society, no uniform practical system could be introduced.

The general land tenure system that has evolved over many years without much government intervention, but at the same time has had its blessing, recognizes the rights of the individual to collect gum from trees growing on the land he previously cultivated. Tribal lands were divided up into village holdings so that cultivation and gum collection were restricted to people domiciled in that village. The head (Sheik) of the village was, and still is, responsible for the allocation of land to his villagers, and the surplus was left under his trusteeship to be allocated to new adults of his village or newcomers who wished to settle in the village.

Each person tapped his own gum garden which grew on his previously cultivated plot. This is known as "geneina" or garden gum as opposed to "wadi" or forest gum which has grown naturally on village waste land known as "ghifar." Villagers have the right to tap portions of the wadi gum and the remainder is left to the Sheik who often uses it as a source of revenue in recognition of his services to the villagers. He lets it out to strangers from other villages, who do not have gum gardens of their own, for an annual rent ("tugundi") or on a half-share basis.

The land has always been regarded a communal ownership not eligible for sale, although cases of land being given as dowry have been reported (Tippetts 1908). After the death of the land holder, it is handed over to the youngest children of the deceased; elder children get their shares from the unallocated village lands. This is because young children are not capable of opening up natural forests for occupation.

Each family was allowed to cultivate as much land as it was able to from the village surplus and thus, could claim the right over the resulting gum gardens. The system developed into properly organized land tenure, whereby a family owned two or more plots in which crops and gum gardens were alternated. The size of the plot depended on the family size and the area of land its members could cultivate. An average family of five members usually possessed around 40 feddans (17 ha) of land for all purposes (Booth 1965). The plots were clearly demarcated against trespass, which is still regarded as a serious offence in the courts. The plots reverted to communal property in the event that the family left the village to settle elsewhere.

During the past decade or two some major developments have occurred which have had a marked effect on the course of the old, well-established system of land tenure and its institutions. The immense growth of human and animal population, which was unforeseen by Newbold in 1935, has caused serious damage to the land, particularly around permanent water points, where overcultivation and overgrazing are now significant land use features. Accordingly, the bush-fallow system has been disrupted and large areas are now completely devoid of gum trees, which were the chief fertilizer and protective cover of the soil. Drought, which severely hit the northern areas of the gum belt, has caused movement of people from those areas to the central parts of the country where living conditions have remained relatively tolerable. This again has caused additional pressure on the lands which were already near devastation.

Besides gum arabic, the sandy soils of western Sudan are a source of crops such as groundnuts, sesame, kardade, melon seed and dura; in addition cattle, sheep and camels are raised. As such, this area ranks very high in the gross national agricultural production and is an important source of foreign exchange. The socioeconomic problems outlined above, if unsolved, are expected to cause a steady decrease in agricultural and animal production leading to a drop in the country's foreign exchange earnings. It is believed that these problems can only be solved by the adoption of a new land use policy based on carefully planned land tenure in order to provide security of settlement and sustenance of the rural life.

The first element of the new policy should outline the solution to the problem of water provision. It has become evident that digging wells close to each other and providing large amounts of water in any one water yard have been the direct cause of overutilization and degradation of land. Water provision to the sandy soil areas of western Sudan should be compatible with the capabilities of the land. In other words water provision should be an integral part of the whole land use policy and never be regarded as a separate measure. The second point that should be considered is the question of soil protection against erosion which has now become a prominent feature leading to a serious drop in agricultural production. Khogali Mirghani (1973) has clearly explained the drop in the average production of crops in Kordofan sands during the period 1963/64 to 1972/73 in spite of the fact that rainfall has not shown similar trends. Apart from the reasons given by Khogali, the writer is of the opinion that such a drop (from 0.28 ton/feddan in 1963/64 to 0.09 ton/feddan in 1972/73) is also related to soil erosion which is a result of overutilization of the land.

It has been shown by many workers that the bush fallow system based on *Acacia senegal*, is the most practical way of sustaining crop production on the light sandy soils. This system, however, needs to be reviewed on the basis of an agrarian reform which should put an end to discrepancies in land distribution which have arisen in the past. The old system of land tenure which has been evolved by the ordinary cultivators has been good enough for the past when living conditions were relatively simple. The system proposed for future land tenure is based on redistribution of the land among the cultivators so that each farmer has sufficient area over which he might practice his shifting cultivation to ensure sufficient income from gum and crops (Seif el Din 1976). The land allocation should be made in a flexible manner, taking into consideration the factors which have led to the disruption of sustained production. In this way the land will remain government property as stipulated in the Unregistered Land Act of 1970, but at the same time the rights of each individual over a specified piece of land will be recognized. Unlike the present communal onwership, in which an individual tries to take out of the land as much as he can without due care for its wellbeing, the suggested system will enable the farmer to see that his plots are well kept and also the government might find it easier to enforce legislative measures aimed at protection and improvement of the environment.

References

Booth, G.A. 1965. *Study of the Gum Acacia Senegal and the Supply of Other Forest Produce in Relation to Land Use Planning.* Doxiades Associates. Athens.

El Mahdi, S.M.A. 1971. *A Guide to Land Settlement in Sudan Khartoum.* University Press. Khartoum.

Hill, R. 1959. *Egypt in the Sudan (1820-1881).* Oxford University Press. London.

Jackson, J.K. and M.K. Shawki. 1950. Shifting cultivation in the Sudan. Forestry Dept. Mem. No. 2.

Mirghani, K. 1973. A report on some agricultural problems in Kordofan sands with special reference to pests and diseases. Seventh Erkawit Conference. Khor Taggat.

Newbold, D. 1935. Letter No. KnP/I.F.I/34 of 11th July 1935. Kordogan Province Archives.

Nicholls, W. 1913. Gum gardens registration. Kordofan Province Archives.

Seif el Din, A.G. 1969. The natural regeneration of Acacia Senegal (L) Wild. M.Sc. Thesis. University of Khartoum. Khartoum.

Seif el Din, A.G. 1976. The role of *Acacia senegal* in the development of agricultural production. Second Agricultural Conference. Khartoum.

Tippetts 1908. Unpublished minutes of a meeting held on 17th June 1908.

Warburg, W. 1971. *The Sudan Under Wingate.* Frank Cass.London.

Willis 1912. Settlement of gum gardens and rights of tugundi. Kordofan Province Archives.

Tenure of Trees or Tenure of Lands?

H.W.O. Okoth-Ogendo

Department of Public Law, University of Nairobi

Introduction

Land lawyers everywhere will be familiar with the question posed in the title of this presentation; namely, whether there can be tenure of trees apart from the land on which they stand. In English speaking Africa, that question presents a peculiar problem. Whereas students of African property law will have no difficulty in returning an *affirmative* answer to that question, those of imposed English law will, equally, have little difficulty in returning a *negative* one! The problem this raises then is: What answer do we return when the two systems intersect or, worse yet, when (as has happened in most of these countries), the former is replaced by the latter through some variant of a tenure reform program? The concern of this presentation is to outline how this problem arises and what some of its implications for agroforestry development are or might be.

Tenure in African property law

Although the specific attributes of tenure systems in Africa vary a great deal, it is now accepted that they possess certain common characteristics that make generalizations on a continental scale possible and meaningful. The first and most crucial of these relates to the notion of "tenure" itself. This, it is now recognized, *primarily* refers to the relationship between people in given socioeconomic and political contexts. That relationship is what generates and determines the incidents of *rights over land*. A distinction must therefore be made between *control* of land and *access* to it in every society.

Control of land, it is now recognized, is an incident of *sovereignty* (in its *jurisdictional non-proprietary* sense) and therefore resides in the political authority of a given society or other segment of it. The purpose of attaching control to sovereignty is to ensure the proper administration of the resource and the equitable distribution of access rights over it. The expectation therefore is that *control rights* will be used *firstly* to guarantee access and *secondly* to respond to the changing needs of society. In this latter sense control rights entail a power of distribution and redistribution as society adjusts to changing claims upon those resources.

Access to land, however, is an incident of *membership* in a society. The nature and quantum of *access rights* any member of society can have will therefore be determined by the network of social relations in which that member participates. This further means that one cannot, therefore, have *general* access to resources. The rights of access, on the contrary, becomes a multiple phenomenon that varies in nature and content with the values and needs society assigns to that member. Thus in any given community, a number of members could each hold access rights expressing a different range or variety of functions, e.g. cultivation, grazing, transit, etc., over the same parcel of land.

The result of this distinction is that the notion of *ownership*, i.e. the appropriation of all rights of use, abuse and disposition of specific parcels of land, is alien to African property law. And this is where most studies of African tenure relations have gone drastically wrong.

The second characteristic relates to the meaning attached to *land* itself. Throughout Africa, the legal conception of land as a resource is confined to the physical *solum* only. This much is evident in the clear separation in indigenous law of the soil from buildings, trees, and other objects affixed to the soil which are *generally* regarded as the property of the party the product of whose labor it is, rather than of the authority in whom control rights are vested or the person entitled to *immediate* access to the use of the soil. This position would also seem to apply to systems of property based on Islamic law.

On account of both or either of these two characteristics one can easily see why African property law would have no difficulty in admitting of tenure of trees apart from the soil on which they stand. One's right to plant, use or dispose of trees or any particular tree would relate not to ownership or even control of the soil but rather to the *nature of access rights* assigned by the community to that member. What this means is also that whether any particular tree can be planted (by whom and where), harvested (and in what manner) or disposed of (and in what circumstances) remains a function of those same relations that generate individual or collective access rights.

Tenure in imposed English property law

Therein lies the fundamental distinction between African and imposed English property law. For not only does English property law admit of *ownership* of resources, such ownership is conceived of basically in nucleated (or individualized) terms. To own, therefore, is to *exclude* any possibility of others having simultaneous rights of the same or different quality over particular resources.

That contrast is dramatically exhibited when that conception of ownership is applied to land. For the legal conception of land in English property law is summarized in the maxim *cujus est solum, ejus est usque ad coelum ad inferos* (whose is the soil, his is also that which is above and below it) As a general rule, therefore, there cannot be tenure of *trees* apart from the land on which they stand. The feudal origins of this rule can hardly be concealed. For in it is also embodied the division of things which a *villein* (i.e. serf) could or could not remove from the soil. And, although English law did develop several exceptions to it through statute and judicial pronouncements, the system of property law which Britain imposed on colonized Africa retained the basic elements of that feudal rule.

The tenure implications of imposition

When first imposed, English property law was restricted only to immigrant communities. As such, a dual system of property law was the general order in colonized Africa. But as part of the colonization process also involved the systematic sabotage of African values and institutions, this *duality* was often more apparent than real. Recent studies indicate that by the end of the colonial era in most of these countries indigenous systems of law in general and of *property law* in particular, had become saturated by key concepts, doctrines and values drawn from imposed English law. Indeed it has even been argued that the concept of *ownership* itself had become part of African property law.

The immediate *tenure* implication of this was that a great deal of normative confusion was introduced into African land relations. And it was that confusion which partly explains why these relations were, in the latter part of the colonial period, characterized by so many protracted disputes over the physical *solum* itself, its products and even specific boundaries between parcels of it. And as long as that confusion remained, it was never quite so easy to determine with clarity who, at any point in time, was entitled to *what rights in what land* and *what rights in what tree*.

As part of an attempt to remove that confusion, a number of African countries (Kenya included) have resorted to tenure reform programs based on the replacement of African tenure systems with structures and norms founded on English property law. What this has generally meant is the individualization of land ownership through a three-tier process involving the adjudication of "*rights amounting to ownership*", the consolidation of parcels, and the registration of titles in a state-maintained, Torrens-based register. The expectation, generally, is that not only would rights to land be clarified, but also that registered proprietors would become innovative, productive and extrepreneurial.

An assessment of Kenya's experience with tenure reform along these lines indicates that further confusion, rather than clarity, may have been introduced into land relations. The tenure reform law prescribes, *inter alia*, that the rights of registered proprietors are *absolute* and "not liable to be defeated" except in accordance with the Act; whereas the social conception of land among rural Kenyans, namely, that it is a collective and functionally heterogeneous asset which must be preserved for the benefit of present and future generations, does not appear to have radically changed. As a matter of fact a resurgence of litigation over who holds what interest in what land is much in evidence, particularly in areas of severe land pressure such as Central Province, Kisii and Kakamega Districts. This development has been given impetus by judicial attitudes which now tend to regard the tenure reform program merely as an exercise in the regionalization of *land administration* rather than in the *privatization* of rights. Indeed the prevailing view in Kenya's High Court now appears to be that the adjudication, consolidation and registration of any land formerly held under indigenous law, cannot free it from the community trust obligations inherent in that law.

The main consequence of this development has been a great deal of *insecurity of tenure* for both registered and unregistered claimants to land throughout the country. This has been further aggravated by a recent law called the Magistrates' Jurisdiction (Amendment) Act, 1983 which provides that all disputes relating *inter alia*, to "beneficial occupation" of land (other than title) must now be taken to the *elders* and not the courts for adjudication. What *beneficial* occupation means, how this is to be determined, and the *corpus* of law with reference to which this is to be done, are not, however, specified in the Act.

In Kenya, in the ten-mile coastal strip where the British colonial administration first extended a property regime based on English law to local communities, but where, in addition, indigenous precepts of law have over the centuries completely coalesced with those of Islamic law, this insecurity is all the more visible and problematic. Although it has been the law since 1908 that registered proprietors of land (mainly of Arab extraction) are entitled to appropriate not only the soil but also all objects embedded in the soil and especially permanent trees such as cashew nuts, coconuts and citrus fruits, it has equally been the firm belief of the vast majority of the largely unregistered (hence technically squatter) population (mainly of African extraction) resident there that those trees and other fixtures belong to them because, invariably, it was they who planted or otherwise affixed them to the soil. In consequence disputes over land and/or fixtures have assumed not merely a proprietary dimension, but also a sociocultural and political one.

Implications for agroforestry

Tenure insecurity arising from confusion in property rights generated by conflict between antagonistic property regimes or value systems can have important implications for land use. The fact that tenure insecurity is itself a reflection of normative instability in society often means that innovative land uses are unlikely to be tried or can no longer be introduced for lack of appropriate units of social organization. The litigation process that constantly accompanies such insecurity in a winner-take-all system further weakens social structures. And, where this generates landlessness and migration, it invariably leads to pressure on public lands (whether used or not) or systematic encroachment on unused or underutilized private lands.

This situation can be very bad for agroforestry, for it may inhibit completely the introduction and development of particular tree species. This is especially likely with regard to permanent trees or those species the planting of which is culturally interpreted as evidence of intention to lay claim to the physical *solum* itself. But more often than not the existing stock of trees could disappear as a result either of wanton destruction or the conversion of land to more immediate, usually agricultural or pastoral uses. The cost or restoring the "forest estate" *per se* or in the form of extensive agroforestry programs, in those circumstances, will then include not merely the technical act of planting trees, but the more difficult social process of reconstructing new norms and institutions.

Tenure and Agroforestry Potentials in India

W.R. Bentley, G.B. Singh and N. Chatterjee

Winrock International, Indian Council of Agricultural Research, and University of California - Berkeley

Introduction

India's rural population is about 570 million people, or 78 percent of its total. The rural land resources include 266 million hectares, of which 143 million are cultivated and 123 million uncultivated. The rural poor are disproportionately dependent on the "cultivated half" (Romm 1981, Johda 1983), and they own or operate the marginal third of cultivated lands, which generally are the 40-50 million hectares of rainfed upland fields.

Land degradation is quite common and serious in India. General estimates of degradation suggest that 150 to 175 million hectares--56 to 65 percent of rural India--are seriously degraded, regardless of cause or the potential for rehabilitation (Bentley 1985). Using the definition of wasteland status as 20 percent or less of annual potential biomass production, Bhumbla and Khare (1985) estimate the wastelands to include:

Cause	Million Hectares
Saline and alkaline	7.16
Wind eroded	12.92
Water eroded -- uncultivated	35.38
Water eroded -- cultivated	38.21
Total non-forest	93.67
Legal forest (very approximate)	25.00
TOTAL	118.69

The bulk of the wastelands are marginal agricultural lands and pastoral common lands, so the impact of productivity loss falls disproportionately on the poor. The same social impact probably is true of the legal forests that are seriously degraded. Grazing and common property or common-access property rights are critical to the causal forces behind most cases of degradation (Bentley 1984).

Land use and tenure problems

The bulk of South Asia was forested at one time (Warner 1982), perhaps as high as 85 percent forest and scrub jungle at the time of Emperor Ashoka (ca. 200 BC). Consequently, most of the lands that are now under cultivation have an ecological potential for woody plants. The current land use and tenure problems can be understood in terms of legal forests, marginal cultivated lands, and commons or revenue lands.

Legal forests

Seventy-five million hectares (Mha) of India, almost one-quarter of the nation, are legally forests. There are three categories:

Reserve forests	39 Mha
Protected forests	23 Mha
Unclassified forests	13 Mha
	75 Mha

The reserve forests were created after the 1857 Mutiny when railroad construction and fuel needs headed the demand for forest resources. The British brought in German foresters who combined the feudal forest traditions of central Europe with not dissimilar traditions of Mughal South Asia (Richards and McAlpin 1983, Tucker 1983). With some notable exceptions, local forest use rights were abrogated.

Much of the forest lands remained under the control of local zamindars (landlords) and the princely states. These lands came under to state control soon after independence, but usually with local rights more or less intact. As this reform was anticipated, substantial areas were cleared and converted to croplands (20-40 Mha) or simply exploited. The lands, whether or not actually forested, were legally viewed as forests. They became protected or, in some cases, reserved forest lands after classification of rights under provision of state land reform acts in the period 1952 to 1957. Unclassified forests still exist in the areas where rights or status have yet to be legally formalized. The forestry departments have never established the degree of use control that the zamindars once held over these lands.

The common tenurial trait of all legal forest lands is that the land is owned by the state. On the whole, the productivity is low, averaging between 0.5 and 0.67 cu.m./ha/yr. Such low growth rates suggest that much of the area has no trees, much is understocked, and some old-growth forests are well beyond biological and economic criteria for optimal harvest age (Bentley 1984).

Agroforestry systems are common in some legal forests in the northeast, especially taungya, and opportunities exist for improving their productivity (Singh 1979). While not yet generally recognized as an appropriate set of land use technologies, agroforestry systems could be used in a sequence of rehabilitative steps or in target systems designed to be more compatible with local people's needs.

Successful agroforestry schemes on legal forest lands will require that rights to tree or grass produce be separated from land ownership. These rights must be adequately secure to entice investment of labor and capital. Experience in India and Bangladesh suggests that an increase in security can be subtle and have strikingly strong results (e.g. the superb response of resettled landless people on protected forests of the Chittagong area with fodder, fuelwood and other trees with minimal assurances of long-term tenure).

Marginal cultivated lands

Much of the marginal cropland area was forested as recently as 35 to 70 years ago (Tewari 1984). This explains the low productivity of the land in aggregate, which is scrub jungle and consequently less well protected. Unless secure irrigation water can be developed, little can be done to make this land more productive in terms of food crop yields. The already low soil productivity is dropping rapidly due to land uses that must be considered inappropriate, given the slopes and soil types.

The tenure rights of farmers vary on the marginal agricultural lands. Many have, in essence, fee-simple rights. In such cases the critical bottlenecks for agroforestry will be access to good information, credit to cover both investment and consumption needs over the transition from annual crops to tree crops, marketing services for various tree products, and similar social infrastructure. Local and intermediary organizations can develop these support systems, often in collaboration rather than conflict with government (see for example Society for Promotion of Wasteland Development and Bharatiya Agro-Industries Foundation 1985).

A significant proportion of resource-poor farmers have their tenure rights clouded by financial and other relationships to the more wealthy and powerful local interests. As a consequence, both parties with interests in these farmlands perceive increased risks, which make agroforestry or other investment prospects appear less attractive even when the real rates of returns are high. Resolving these difficulties requires alteration of the basic structure of cultural relationships among classes and castes, which most likely will create local social instability for a period. When changes do occur, there may be a tendency, if agroforestry or similar land use technologies are financially attractive, to displace the debtholders in favor of some other form of tenurial holding.

Even if tenure rights in general are reasonably secure and comprehensive, traditional fallow season grazing rights can affect the viability of agroforestry schemes where the tree crops are palatable to livestock. This problem is faced in any area that shifts from rainfed to irrigated agriculture, but the perception of trees as a crop that deserves protection by animal owners is slower to develop. Some of the solutions developed for management of the common pastures may be applicable.

Commons or revenue lands

About 48 million hectares of community lands, along with legal forest, make up the uncultivated half of India. Rights of one sort or another are held on these lands, sometimes by whole communities, sometimes by clans or tribes, and occasionally by individuals. Legally the control of such lands resides with the state revenue departments; hence, the name revenue lands. Where no one person holds rights, the rights are obscure, and where political forces favor abrogation of existing rights, revenue lands may be assigned to specific individuals (e.g. landless families) or communities. This was the case in the well known West Bengal social forestry scheme that gave many families a hectare of land and subsidized their rehabilitation of the area with trees that they owned. In general, however, these lands are treated as commons and their productivity is low and declining. No agency is in charge of regulating use or making investments, except where a Panchayat or equivalent village unit has gained strong control.

Social fencing is an example of the kind of device that can be used by villagers to control use and create an environment suitable for investment in increased productivity. Basically, a village association creates peer pressure to keep animals off grazing areas during certain seasons or all the time, and often fodder cutting with stall feeding replaces grazing. In many areas, this practice alone leads to substantial increases in grass fodder productivity over a year or two, as root systems fill out and the total site is covered by vegetation (i.e. maximum photosynthetic capacity is achieved). Investment in seeding or planting of grasses and fodder trees can follow once a village controls its own members and protects against trespass by outsiders. The ease of achieving these results depends on many local cultural variables, but especially on the degree of social homogeneity and the value of the resources in question.

Social fencing has been used most successfully in cases where either the village already has strong traditions of community management or was able to displace a fodder contractor on protected forest lands. The village's success in managing its own grazing commons is a useful experience in taking on other social forestry activities, and opens up the possibility of non-private agroforestry systems.

Discussion and summary

Two striking features of India are of concern in regard to the relationship between land tenure and agroforestry potential. First, there are tenurial insecurities on at least 100 million hectares of cultivated and common lands, and the forestry department control appears minimal on at least 30 to 40 million hectares of legal forests. Second, these lands often have a forest or scrub jungle ecological climax, so there are reasonable expectations of increased biological productivity and reduced soil degradation by shifts toward trees as part of the managed vegetative system. Coupled with rising tree product prices, falling agricultural commodity prices and rising land management costs in real terms (Chatterjee 1984), agroforestry land use technologies have social and economic potential.

Clarification and assignment of tenure rights could increase productivity. Assignment of rights to villages, organizations and individuals that represent the resource-poor or landless offers potential for using the productivity increases to alleviate poverty. The actual potential for doing this in specific cases, however, depends upon how entrenched the local power structure is and how the elite view the advancement of people who are below them in social and economic status.

References

Bentley, W.R. 1984. The uncultivated half in India: problems and some solutions. Ford Foundation Discussion Paper No. 12. Ford Foundation. New Delhi.

Bentley, W.R. 1985. Agroforestry--a strategy for research and action in India. Ford Foundation Discussion Paper No. 17. Ford Foundation. New Delhi.

Bhumbla, D.R. and A. Khare. 1985. Estimate of wastelands in India. Society for Promotion of Wastelands Development. New Delhi.

Chatterjee, N. 1984. Economic aspects of social forestry in India. Paper for FAO/EWC Workshop on the Socio-economic Aspects of Community/Social Forestry in the Asia-Pacific Region. Bangkok.

Johda, N.S. 1983. Market forces and erosion of common property resources. Paper for ICRISAT International Workshop on Agricultural Markets in the Semi-arid Tropics. Patancheru, Andra Pradesh.

Richards, J.F. and M.B. McAlpin. 1983. Cotton cultivating and land clearing in the Bombay Deccan and Karnatak: 1818-1920. In R.P. Tucker and J.F. Richards (eds). *Global Deforestation and the Nineteenth Century World Economy.* Duke University Press. Durham, North Carolina.

Romm, J. 1981. The uncultivated half of India. Parts I and II. *Indian Forester* 107: 1-23, 69-85.

Singh, G.B. 1979. Agro-forestry in eastern Himalayas. In *Proceedings of the Imphal Agro-Forestry Seminar.*

Society for Promotion of Wasteland Development and Bharatiya Agro-Industries Foundation. 1985. Utilization of wastelands for bioenergy. Summary program for workshop. Pune.

Tewari, K.M. 1984. Energizing social forestry. Presidential address ISTS Satellite Seminar on Agroforestry Systems. Ranchi.

Tucker, R.P. 1983. The British colonial system and the forests of the western Himalayas, 1815-1914. In R.P. Tucker and J.F. Richards (eds). *Global Deforestation and the Nineteenth Century World Economy.* Duke University Press. Durham, North Carolina.

Warner, F. 1982. Indo-Swedish forestry programme II 1982/83 to 1986/87. Background Document. Swedish Embassy, Development Cooperation Office. New Delhi.

Some Tenurial Aspects of Environmental Problems in Nepal

Kk. Panday

International Centre for Integrated Mountain Development, Kathmandu, Nepal

Introduction

The Hindu Kush Himalaya region is characterised by its ecological diversity. Between the tropical jungles of Burma and the arid mountains of Afghanistan lie the greatest mountains and deepest gorges of the world. Here live almost 10 percent of the world's population. One of the most common conditions of these people is the poverty they live in. They also show the limit of man's patience and ingenuity.

The most common socioeconomic features of the region are: 1) rising population (Nepal alone doubled its population over the last 30 years), 2) poverty related to the subsistence farming in the fragile environment, 3) a strong will to live in this hostile environment and carve out an existence.

The most alarming situation at the present time is the deforestation of the mountains and loss of soil fertility.

The founder of Buddhism, Sidhartha Gautam, while recognizing the problem of fuelwood, fodder and timber, asked his people to plant a tree every five years. The problem 2500 years later is such that every individual needs to plant at least 10 trees every year for over 10 years if Buddha's land is to reverse the process of destruction.

There are conflicting views on the causes and effects of deforestation and the population increase. It may be wrong to blame the rise in population as the primary cause of deforestation. There are certain other developments that need to be looked into in the recent history of the country.

The process of large scale logging may have been initiated by the Nepalese kings of the 9th century when they incorporated artistic wood carvings in their elaborate constructions. The temples and dwellings built in centuries past are still witness to that development.

It is of interest that the history and pattern of forest use and deforestation in the northeastern hill districts of Nepal has been wholly determined by factors arising within Nepal rather than extraneously (Mahat et al. 1984). But the depletion of forests in the plains of southern Nepal was primarily caused by the opening of the hardwood forests for logging to satisfy large external markets, as well as internal land hunger and construction demands of the 1950s and 60s. If we look back we observe the last three decades as the period of greatest importance from the viewpoint of environmental concerns.

A period of confusion

Nationalization of forests

The autocratic rulers prior to 1947 regarded forests as their private property. Forests of good chirpine trees were felled and lavish palaces were built. When political change came in 1947-51 the new government asserted its authority by declaring all forests nationalized in 1956 during the process of democratization. Those who had forests on private lands lost that part of their holdings.

In 1961 the owning of private forest plots was allowed but a ceiling was determined: 1.25 ha for the hill areas and 3.5 ha for the plains of the south. But the formerly nationalized forests were not returned to their owners. It went even further and all the land adjoining the forest area and left fallow for two years were defined as national forest land (Manandhar 1982).

The change in the authority of the state between 1947 and 1957 was from an autocratic to an unstable democratic system. The nationalization of forests occurred when the government needed to assert authority but it happened, as it now seems, without any basic understanding of the implication of such acts. By the time government became more or less stable and people-oriented the damage had been done.

Land reforms

In 1964 the government enacted a land holding ceiling. The total land holding permitted to the hill people was about 4.5 ha per family. The average size of the farm family is 6-7 persons. This implied that there would hardly be 3.25 ha of land left for cultivation (after conservation measures) to feed, clothe and shelter the family.

However, an analysis of independent data, from West Central Nepal, indicates that each family of 5 to 6 members with an average farm holding of 1.25 ha requires, with present agricultural and forestry management practices, an additional 3.5 ha of land for fodder, 0.3 to 0.6 ha for fuel and 0.4 ha for timber to sustain current activity (Wyatt-Smith 1982): more, in other words, than the private cultivated land holding permitted by the law. In most cases so much land area is not available in the forest for private use.

In a subsistence economy with the fragile environment of Nepal, whose population has a doubling time of 30 years (by the year 2000 there will be 24 million inhabitants in the country), the present land use systems themselves pose a major constraint to both food production and the preservation of forest land.

Period of destruction

It is believed that the Nationalization Act of 1956 has hastened the process of forest depletion, especially in the hills and the mountains (Manandhar 1982). The exception observed by the people was to spare the forest plots around holy shrines, which were not managed but also not damaged. Kathmandu Valley alone can count 5 of these ancient forest plots despite high land pressure.

There had been traditional management systems which controlled forestry resources and ensured that the community needs for forest products were met without damaging the resource base. These systems collapsed as soon as the Nationalization Act declared them illegal.

Period of learning and assessing the damage done

The period of assessment and understanding of the deforestation problems took 10 agonizing years (1967-1977). In 1978 new community and conservation oriented regulations came into effect. There had been no provision for 20 years (1957-1977) to provide private or community ownership of the forests that would motivate people to plant trees. The new regulations made it possible for the government to turn over government owned forests and lands to community ownership as a direct incentive to development and conservation efforts at the local level.

In spite of the new laws the political will to save the remaining forests was still lacking, as manifested during the political upheaval of 1980. It was later painfully realized that damage was again done to the forests, even after new laws had come into effect. The forests are being severely exploited at an increasing rate partly due to the perception that the government owned forests belong to nobody.

Now, again, the impact of the newest laws on the people needs to be looked into. They have certainly brought some positive reaction at the community ownership and management level. However, the response at the individual level is important if Nepal is to check the environmental problem which may have deep roots in the land and forest tenure systems. The present land ceiling is just not high enough to motivate individual farmers to plant trees and create forests.

The farmers already do their best in their present situations. Allocating separate plots of land within the holding for afforestation is not possible because of size constraints. Also, since there is no market in the hills for timber, convincing the farmers that planting trees would bring more income than farming is unlikely. The planting of trees traditionally used for fodder may be more attractive to the farmers, particularly if done in non-competitive agroforestry arrangements. Tenure concerns may be an incentive to this type of tree planting. Many farmers believed that they could save their land from nationalization in 1956 because they had planted fodder trees only very sparsely.

How has China coped with a situation similar to Nepal's? Their new policy recognizes the fact that economic development in the mountains is dependent on the rational use of the mountain environment and that it must rely primarily on forestry.

What is interesting in the Chinese model is the fact that people can now sign contracts individually or in groups, for using barren hills and remaining forests. The government issues ownership rights to the citizen(s). They can own the afforested land and manage it by themselves, and their offspring can inherit the land. Not only can they use their trees freely for domestic purposes, but they can also sell a part of the forest product at free market prices. The unwarranted emphasis on food production and self-support in grain production has changed in the mountains of China (Ji 1985), through adoption of an acceptable tree tenure system, in the Chinese way.

References

Ji, X.Y. 1985. Issues of mountain development in China. ICIMOD Memo. Kathmandu.

Mahat, T.B.S. et al. 1984. An historical perspective on the forests of Sindhupalchok and Kabhre Districts of Nepal. Symposium an der Eth. Switzerland.

Manandhar, P.K. 1982. *Introduction to Policy, Legislation and Programmes of Community Forestry Development in Nepal.* HMG/UNDP/FAO Community Forestry Dev. Project. Kathmandu.

Panday, Kk. 1982. *Fodder Trees and Tree Fodder in Nepal.* SDC and FFRI. Bern and Birmensdorf, Switzerland.

Wyatt-Smith, J. 1982. *The Agricultural System in the Hills of Nepal: The Ratio of Agricultural to Forest Land and the Problem of Animal Fodder.* APROSC. Kathmandu.

Land Tenure and Agroforestry in Forest Land in Thailand

Komon Pragtong

Royal Forest Department, Bangkok, Thailand

Introduction

Since 1961, it has been a goal of the national social and economic development plan of Thailand that 50 percent of the national land or 256,000 km^2 be reserved as forest land. The target will be reduced to 40 percent or 200,000 km^2 as the population increases in order to open more land for agriculture (National Economic Development Council Office 1960). To meet this target, forest land has been reserved under the Reserved Forest Law of 1954. In 1982, the forest area was declared to be 20.02 million ha reserved forest, 2.38 million ha national park, 2.03 million ha wildlife protection area and 2.97 million ha of additional forest still to be reserved. The national park and wildlife area will be protected under the national park law and wildlife protection laws respectively, which are more conservation oriented than the reserved forest law.

The opening of forest for agricultural land has taken place rapidly since 1961. According to aerial photo and the LANDSAT studies on the trend of the loss of forest land, 273,628 km^2 or 53.33 percent of the national area was forested in 1961 and this declined to 156,600 km^2 or 30.52 percent in 1982 (Boonchana and Jaruppat 1983). It is estimated that about 117,028 km^2 of forest land or 22.81 percent of the national land has been cleared for agriculture during the last two decades. Rapid population growth after the Second World War, together with low-productivity traditional farming techniques, created the need for more agricultural land, including more land for export crops like tapioca, sugar cane and maize (Dulyapach and Pragtong 1984).

Land tenure issues in forest land

Reserved forest land is considered government land and is protected under the National Reserved Forest Law (1954). As provided by the law, individuals can get legal land tenure in the forest in 3 cases: 1) one may ask permission to utilize the forest land temporarily for particular purposes and under specific obligations stated in the permission paper; 2) one may ask permission to do research and technical work in the forest land; and 3) one may ask permission to carry out large scale forest plantation in deteriorated forest land for a certain period of time and under specific obligations as stated in the permission paper (Phuphae 1985). However, because of the complicated process of seeking and being granted permission, the needs of a great number of clients are not met soon enough. To cope with the problem authority to grant permission has just been delegated to the regional level from the ministry level. Some farmers claim more land than is possible under the regulations, adding to a difficult situation.

Illegal land tenure in forest land has been accepted among the farmers. A natural boundary like a canal, hill or tree is used as the land mark. As the farmers come from the same town or have the same origins or go to the same local creditors, the land boundary is known and accepted. Moreover, those who harvest the land pay land use tax to the local headman who forwards it to the district land authorities, whether the harvest is from government land or from private land. This practice sometimes leads to a misunderstanding of legal land tenure in the forest land, since the farmers assume that their right to be there is validated by the acceptance of their tax money.

Since the issue of legal land tenure is unclear to the farmers, they make the land holding rules among themselves and tend to grow cash crops more than permanent crops to serve the available local market. The taungya system can be employed on some lands to harmonize the farmers' cropping system with the government's forest plantations. But there are many disputes concerning land tenure between the farmers and the government's forest plantation officers, which leads to stagnation in the progress of the forest plantation program.

Forest village programs

The Forest Village program may be considered a modified form of the taungya system. Population pressure in Thailand after World War II has played a large role in promoting illegal occupation of forest land. It was estimated in 1978 that there were roughly 6 million people who illegally claimed land in the forest. The taungya system has been employed in those forest plantation areas that have the problem of illegal land tenure. The system was adopted by the Royal Forest Plantation in 1906 and initiated in Prae, Northern Province. In this system the farmer was encouraged to plant cash crops between the forest trees. However, it was found that the taungya system was not successful because in many cases the farmers paid most of their attention to their crops, epecially when crop prices increased, and neglected the forest trees. In addition, the allotted cultivation period was only 2-3 years, so the settled farmers subsequently turned to shifting cultivation. Later, the taungya system was abandoned and the forest plantation was created on a labor-hiring basis for all the activities of planting.

However the pressure of land needed for agriculture still plays a role in the forest plantations. Landless farmers can ask for permission to intercrop in the spaces between the trees on a 2-3 years contract basis. The system, which can be considered another form of modified taungya system, has been adopted widely in the planting unit of the Royal Forest Department. However, the intercropping system has not proved fully successful since in some areas the farmer who used to harvest the land has not been satisfied with it.

In 1967, the Forest Industry Organization (FIO), the government enterprise of Thailand initiated the project called the Forest Village System of plantation. The system is aimed at settling landless people in communities and providing a source of labor for plantation work. Each village normally is comprised of 100 families. Each family is allotted 1.6 ha of land per year for clearing, planting crops and forest trees, and 0.16 ha of land for a dwelling and homegarden. In the village, electricity, water supply, school, temple and other amenities are provided. A reward of US$5 is given to each family for every 0.16 ha of trees planted. In addition, if any family works on clearing land and planting trees for 3 consecutive years and on an area over 4.8 ha it will be paid a bonus of US$ 75. A quota of two laborers from each family is recruited to work for the plantation. Such employment adds substantially to the income of the family. Under this scheme, each village will establish 160 ha of plantation a year. The area of a forest village unit should cover the whole rotation of the species planted (Corvanich 1974). There were 41 forest villages in the plantation program of FIO in 1982.

In 1975, to cope with the pressure of land use, the Royal Forest Department adopted the Forest Village approach in their plantation program with some modifications. The cleared forest area is subdivided into 2 zones. The first zone is the watershed which is designated unsuitable for agriculture. The second zone is the land suitable for agriculture. The forest land in the first zone is on prohibited land where the conventional plantation program has to be implemented. Forest land in the second zone is set aside for the Forest Village system. The village is laid out for 150 families. Landless farmers in the watershed or farmers in areas unsuitable for agriculture will be resettled in the village. Not more than 2.4 ha of land in the second zone will be provided to each household. The area will be divided into the housing area and the cultivated area. In the housing area the farmers will be grouped as a village where 0.16 ha for housing and backyard gardening is made available.

The cultivated area of 2.24 ha will be used for individual permanent cropping. Members of the village have the opportunity to work in the forest plantation project surrounding the village on a hired basis. Those who would like to grow crops between the trees can apply on a contractual basis. The plantation work is to assist the farmers in getting established during their first five years in the village (Royal Forest Department 1982). Electricity, water, schools, a village temple and other amenities are also provided in the forest village. There were 144 villages all over Thailand in 1984 (National Forest Land Management Division 1985).

There is not much difference between the two Forest Village systems. However, the farmers in the FIO system will earn their livelihood on the plantation which provides land for crops. The land and tree tenure rights rest temporarily with FIO as a concession of the RFD. The FIO obtains the legal land tenure for the planting site under the provision of the National Forest Reserved Law (1954) and the farmer will have crop tenure at the tree planting site. In the RFD system, the farmers will get land tenure in the land provided to them for their own use. The land cannot be sold but may be transferred through wills. The crop tenure in the tree planting sites belongs to the farmer on a contractual basis.

The S.T.K. land certificate program

As there are a great number of farmers in the forest land, the Forest Village program cannot cope with the whole of the forest land tenure problem. Moreover, the Forest Village program has been labeled a high cost project because of its integrated nature. Also the Forest Village is a means to carry out large scale forest plantation and it has to be attached to a forest plantation project of the government. Problems of forest land tenure not addressed by the Forest Village program have called forth a new program, the so-called S.T.K. (Sit Thi Thamkin or "Right to Harvest") Land Certificate program.

The S.T.K. Land Certificate program has been promoted since 1979. The program is designed to accept and work with the actual land holding of the farmers in the forest land. The deteriorated forest is divided into 2 zones, the upper watershed area and the land suitable for agriculture. The upper watershed area is protected while the land suitable for agriculture is offered to individuals through S.T.K. Land Certificates. The certificates for each individual will be for not more than 2.4 ha. Those who have claimed more than 2.4 ha will be permitted to keep this land on a temporary basis, as provided by the Reserved Forest Law of 1957. Under the program, the actual land tenure system in operation locally will be accepted by the government. The tree planting program has been encouraged for those farmers who hold more than 2.4 ha. Twenty percent of the extra land will be planted with free seedlings provided by the government. The land and tree tenure will belong to the farmers and can be transferred through wills but cannot be sold (Royal Forest Department 1981). As of 1984 607,945 ha of forest land has been offered to 366,517 farmers under the S.T.K. Land Certificate program (S.T.K. Office 1985).

Agroforestry research needed

Intensive agroforestry land use systems have been practiced by those farmers who live in the Forest Villages. Agrisilviculture and silvopastoral systems have been employed in most of the forest plantation projects of the government. Moreover agroforestry has been practiced widely in the homegardens of farmers in the lowland areas of the central plains. The agroforestry system has great potential for adoption by the small landholder who can obtain legal land tenure in forest land.

Research in agroforestry has been carried out in upland government forest plantations, where the intercropping of trees and crops has been studied. There are six research units located in different topographic zones of the uplands. In the highlands, there are two research units studying the adoptability of the intercropping of fruit trees in highland areas. Most of the research projects are aimed at providing basic data on the intercropping patterns practiced in the government forest plantations. Thus far, there has been no real attempt to transfer the technical know how to the farmers (Pragtong et al. 1984).

The knowledge of intensive farming through agroforestry can be transferred to those farmers who have legal land tenure in the forest land, but to do this the agroforestry extension program must be strengthened. In the extension program many tasks should be included. First, there should be research on the tree and crop combinations and patterns of farm management which will be best suited to the local socioeconomic and ecological factors of the various regions. Second, there should be a means to assemble the agroforestry data available from various authorities who have worked in the private lands outside the forest area. A central body on agroforestry should be formed to take the role of coordinator of agroforestry research.

Third, as the research and development must answer local needs, the farmers, who will be the clients for technology transfer, should attempt to form a farmers organization at the community level. The institution will deal with the mechanics of technology transference and participation in development. The farmers have to build up their own capability to deal with new technologies. Finally, a training and assimilation program should be carried out for better understanding between the farmers and the officers who will use the agroforestry system in the forest land.

References

Boonchana, K. and T.C. Jaruppat. 1983. Situation of the forest in Thailand over a period of 21 years. In *Proceedings of the Annual National Forest Congress-1983*. Royal Forest Department. Bangkok.

Corvanich, Amnuay. 1974. Forest Industry Organization's Forest Village. Technical Paper. Forest Industry Organization. Bangkok.

Dulyapach, S. and K. Pragtong. 1984. Balancing of national agricultural land use and forest area with special emphasis on future trend: A workshop report on the problem of lands with declining and stagnating productivity. 6-9 November 1984. Pataya.

Forest Industry Organization. 1983. *Annual Progress Report 1983*. Forest Industry Organization. Bangkok.

National Economic Development Council Office. 1960. *The National Economic Development Plan 1961-1966*. Sassana Print Co. Bangkok.

National Forest Land Management Division. 1985. *Progress Report on the Results of the Forest Village Project (1976-1985)*. Royal Forest Department. Bangkok.

Phuphae, S. 1985. *Land Use Permission in Forest Reserves*. Royal Forest Department. Bangkok.

Pragtong K., M. Phothitai, S. Bangliang, and M. Jamroenprucksa. 1984. Modified taungya system: agroforestry in Thailand. Country Report to the ICRAF Course on Agroforestry held at the University Pertanian, Malaysia. October 1-20. Serdang.

Royal Forest Department. 1981. *Working Procedure on S.T.K. Land Certificate Project*. Royal Forest Department. Bangkok.

Royal Forest Department. 1982. *Forest Reserve Rehabilitation Project*. Royal Forest Department. Bankgkok.

S.T.K. Office. 1985. *Progress Report on the Results of the S.T.K. Land Certificate Project 1982-1985*. Royal Forest Department. Bangkok.

Notes on Tenure and Agroforestry in Indonesia

Sediono M.P. Tjondronegoro

Department of Rural Sociology and Economics, Bogor

General statistics

The Republic of Indonesia is the world's largest archipelago, covering approximately 1.9 million square kilometres, and contains several large islands, such as Sumatra (473,606 km^2), Kalimantan (539,460 km^2) and Irian Jaya (421,981 km^2), where forest areas are still vast.

A recent tentative survey by the Department of Forestry indicated that some 143 million hectares (75 percent) are covered by forest land, among which:

34 million ha is protected forest
1.8 million ha is recreation forest
64 million ha is production forest
30 million ha is conservation forest

It should be borne in mind, however, that an earlier account of forest lands (Central Bureau of Statistics 1982) has shown lower figures for some, and higher figures for other categories referred to above. Some other categories such as mangrove forest, swamp forest and coastal or peat forest may account for the differences.

Roughly speaking production forest covers 50 percent of the estimated total forest area, and reserved forest approximately 24 percent, and this area has been marked as government controlled land since the nineteenth century.

Agrarian reform

Seen in a historical perspective, the following measures pertaining to agrarian reform have occurred in the 20th century.

First: The Cadastral Registration (1811-1816) under S. Raffles, especially in Java, which paved the way for the introduction of the Cultivation System (forced cultivation of exportable crops) in 1830.

Second: Agrarian legislation through which the Dutch colonial government declared all land to which title could not be proven, as government controlled land (*Domein Verklaring*). One of the long term consequences was that much of the upland, both on Java and the large islands of Indonesia, were kept under government supervision. Long term leases up to 80 or more years were given to private companies, most of foreign origin, to establish large agricultural estates for export crops such as rubber, tea, cocoa and oil palm.

Third: National legislation of a Comprehensive Basic Agrarian Reform (1960) and a Law on Tenancy and Share cropping.

Two distinct patterns of agrarian conditions and of agricultural practices eventually developed in densely populated Java and Bali, and the so-called "Outer Islands." In the former the population has a strong preference for cultivation of food crops, while in the latter shifting cultivation and forest utilization (rubber tapping, gathering of rattan and other forest products) have been prevalent.

Yet rapid population growth all over Indonesia after the Second World War as well as increasing forest concessions in Sumatra, Kalimantan and Irian Jaya have caused an annual expansion of "critical land," i.e. forest land, opened and left idle afterwards.

A recent survey by the Department of Forestry indicated that today critical land has reached a frightening extent of 43 million hectares, among which 20 million hectares are within and 23 million hectares are outside the forest area. Annually about one million hectares are added to the critical land surface due to shifting cultivation in the forest area, mostly on the larger islands of the country.

The expansion of critical land due to shifting cultivation, as well as to careless forest concessionaires, has also had a serious impact on the existing 81 watershed management areas all over the country, of which now some 36 are in critical condition and 22 are under rehabilitation.

Therefore between 1970-1981 reforestation and regreening or afforestation areas have increased from year to year, reflecting the need for more rapid rehabilitation of deteriorating forest areas (Central Bureau of Statistics 1982). Reforestation increased 983 percent and regreening increased 838 percent which would mean an average annual increase of 98.3 percent and 83.8 percent, respectively. More intensive rehabilitation programs are carried out in densely populated areas of Java, where the urgency is considerably higher.

But the regreening programs have suffered problems and bottlenecks in the way of poor coordination, particularly since such programs often cover a number of adjacent administrative areas and are implemented by various sectoral departments. The concept of setting up an "authority" has often been rejected since it would interfere with local administration.

Ecological zones and sectoral concerns

The ecological zones on the island of Java could be delineated or visualized by means of an idealized transect:

- brackish water fish ponds along the coast, sometimes in combination with mangrove or swamp forest;
- foodcrops on the lower peneplain, not infrequently in combination with technical irrigation;
- higher up, upland food crop agriculture, animal husbandry and tree crops;
- forest land from approximately 600 m or higher along a mountain range of young and old volcanoes running through Java's central region.

The ecological zones of Sumatra follow somewhat the same delineation as above, except that the mountain range is along the west coast, leaving little possibility for lowland food crop cultivation between the mountain and the coastline. The broader peneplain with vast areas of swampy land is to the east, where the relatively low population densities still allow shifting (or swidden) cultivation. More sedentary food crop agriculture is found along the big rivers, where tidal movements can be used for wet rice cultivation (*sawah nyapah* and *pasang surut*).

The vast forest areas still available can be utilized to derive off-farm income from and allow cultivation of perennial cash crops such as coffee, pepper and rubber, where cycles can be extended 8 to 30 years or even longer.

Ecological zones in Kalimantan and Irian Jaya are perhaps more alike than Sumatra and Java with vast forest areas and a concentration of population in the coastal areas and along the banks of large rivers. The land surface is undulating with vast swampy areas deeper inland. Shifting cultivation is the most prevalent mode of subsistence for the indigenous population; however, it has become extremely confined, due to the expansion of forest concessions to logging companies since the 1970s.

Indonesia enacted a Basic Agrarian Reform Law and a Basic Law on Tenancy and Sharecropping; however, there has been no success in formulating detailed guidelines for effective operation, so that in many respects the implementation and enforcement of both laws is extremely slow and weak.

Tenure

Tenurial problems have been significantly more associated with Java and Bali, where the concept of absolute individual ownership has been rooted for close to two centuries now, than with the Outer Islands, where the indigenous population is mainly familiar with usufruct rights, on the communal land of the clan or tribe (*Marga*).

Ownership of land in areas where shifting cultivation previously prevailed was initially introduced by Javanese migrants in 1905. They received land certificates after they proved to be capable of clearing and cultivating the allocated land property.

In cases of spontaneous migration, land acquisition by migrants is negotiated with the local clan heads. However, ceremonial tokens of land allocation may not imply actual transfer of ownership of the land, a concept which is alien to the average shifting cultivator.

Transmigration, the government resettlement program, has acquired land through negotiations between the provincial and regional administrators and the owners, with compensation for the tree crops growing on the acquired land. Not in all cases is it a clear-cut money transaction. Sometimes compensation in the form of infrastructural developments and provision of health and educational facilities to all is also promised. For the indigenous people the idea of a greater labor supply and a better infrastructure through transmigration programs is most welcome, but they have never really accepted the concept of giving away part of their ancestral agricultural or forest land.

Differences in "ownership" concepts have, in new settlement areas, indeed given rise to conflicts about control over land. Population densities and the increasing prices of land under improved cultivation can both be seen as causes of such conflicts or social tensions.

Although national legislation accepts absolute land ownership by individuals, customary law cannot be disregarded immediately; in the period of transition, law enforcement is not always the most appropriate way of solving conflicts about land ownership.

Cases of agroforestry

Implementation of certain forms of agroforestry have been known in Indonesia for a long time, both in areas of shifting cultivation and in areas with sedentary cultivation on rather poor upland soils.

The "*Damar mata kucing*" or *Shorea javanica* (Dipterocarpaceae) agroforestry practice in the forests of southern Sumatra, was noted as far back as 1937. The first plantations of *S. javanica* for resin production may be traced to the 19th century. The resin has long been a cash-crop sold throughout the year to local traders (Torquebiau 1984). The small settlement area, usually on the fringe of the agroforest, developed on the hilly parts with rice fields occupying the valleys.

Although the local cultivators may be tapping resin from "individualized" trees, no tenurial relationships with regard to the forest land on which they grow seem to exist. It is a typical shifting cultivators' pattern, with usufruct rights on the land in the hands of the first cultivator. This pattern usually also applies to the rice fields along riverbanks in the valleys. Individualized usufruct rights may be passed on to family members of the younger generation.

The *S. javanica* agroforest may be referred to as a sedentary agricultural practice since the cultivation of rice and resin would commit the cultivators to dwelling for longer periods in smaller villages. Also the fact that, if tapped properly, the damar trees can be tapped considerably longer than 50 years suggests that permanent settlements in small communities is the basic pattern.

The first damar trees may have been tapped in natural forest, but the first "plantations" were probably on improved swidden cultivation (*ladang*) plots, and successfully changed into plantations proper through agrisilvicultural (taungya) techniques.

Another interesting case of plantation development from an indigenous forest area emerged in Abai Siat, West Sumatra. Here, under project conditions, an "individualized" forest area where the tapping of rubber trees in the wild had been predominant was converted into smaller rectangular plots aimed at producing at least equivalent volumes of latex or its money value. The smaller rectangular plots did indeed become at least as productive as the more extensive forest area by means of improved rubber clones, fertilization, etc. They also had the advantage of proximity to the settlement area and to the market center, so that the tapping and sale of resin became much easier.

A nucleous estate was built in the center of the area for purchasing the latex and processing it in the mill. The organization of the plant also provided extension services over and above the selling of fertilizer, pesticides and equipment in a cooperative spirit. It is not clear, what tenurial arrangements might have developed at a later stage of this case of smallholders in the rubber estate, but what was already discernible in the initial stage was a practice with which the shifting cultivator is normally unaccustomed, i.e. share cropping and tenancy relationships. This may have emerged from the immigration of peasants from other areas, attracted by the employment opportunities initially offered by the project. In the long run, the more diligent share croppers may grow economically stronger and acquire some land.

The rejuvenation of teak plantations, carried out in 19th century Java under the Dutch colonial administration, also employed agroforestry techniques. The system of intercropping known as *tumpang sari* was practiced in the forest district of Semarang (Central Java) as early as 1881 and was later applied in other forest areas of Java. A harvest of approximately 800,000 quintals of upland rice per annum was obtained from a reforested area of 40,000 hectares. At that time it was equivalent to 10 percent of Java's total upland rice production.

There is still a discussion going on as to whether or not such a level of production would be sufficient to sustain the population around densely populated, upland forest areas on the island of Java. It was calculated that the use of high yielding varieties and improved techniques of rice cultivation will become feasible if yields exceeding 1.6 ton of dried paddy per hectare can be obtained.

Aside from yields, however, agroforestry intercropping on forest land tries to combine the need of the teak plantation to maintain a labor force and to achieve the highest possible security, with the demand of local people for a living. Forest land is usually government controlled land, often leased for a period of up to one century. In order that the local farming people do not jeopardize the teak plantation by cutting, they are allowed to utilize forest land for intercropping with foodcrops. Some additional farm income was also derived from teak wood leaves for wrapping material at local bazaars, but now these are increasingly replaced by plastic bags.

References

Central Bureau of Statistics. 1982. *Indonesia Statistical Pocketbook*. B.P.S. Jakarta.

Cernea, M. 1985. Units of social organization sustaining alternative forestry development strategies. World Bank. Washington.

Government of Indonesia. 1960. *Undang-Undang Pokok Agraria* (Basic Agrarian Reform Law of Indonesia). Department of Information. Jakarta.

Kartasubrata, J. 1983. Partisipan rakyat dalam pengelolaan dan pemanfaatan hutan di Jawa (Grassroots participation in the management and utilization of forests in Java). I.P.B. Bogor.

Tjondronegoro, S.M.P. and G. Wiradi. 1984. *Dua Abad Penguasaan Tanah* (Two Centuries of Control Over Land). P.T. Gramedia. Jakarta.

Torquebiau, E. 1984. Man-made dipterocarp forest in Sumatra. *Agroforestry Systems* 2: 103-27.

Weiner, J.S. 1972. Tropical ecology and population structure. In G.A. Harrison and A.J. Boyce (eds). *The Structure of Human Populations*. Oxford University Press. Oxford.

Notes on Agroforestry Projects in Developing Countries

Loekman Soetrisno

Rural and Regional Studies Center, Gadjahmada University

Background

In most developing countries, including Indonesia, the main issue in forestry management is how to compromise the two different views of the two groups involved in the forestry management system, i.e. the poor members of the forest based communities and the government.

For the poor members of the forest based communities the forest has two special functions. First, the poor members of the forest based communities view the forest as a solution to the problem that they are constantly facing, i.e. to get a piece of land where they can cultivate food and other crops to meet the consumption needs of their individual families. Second, the forest also functions for the poor as a source of fuel, building materials, and sometimes as a source of extra protein, as the forest is also a hunting ground for them.

On the other hand, the government has different interests in the forest. For the government the forest is a resource to be exploited to generate revenues needed to finance national development. Secondly, the government also believes that forests need to be protected for ecological reasons.

These two conflicting views of the role of forests in human life are not easy to accommodate in the forest management system which is currently being used by most governments in developing countries. Consequently the "forestry world" in developing countries is often colored with conflict between the poor who want free access to the forest and the government apparatus which tries to prevent them.

It is to accommodate these conflicting interests that forestry scholars offered a new concept within the forestry management system, i.e. agroforestry. As a concept, agroforestry is revolutionary, resulting from an honest, courageous, and holistic understanding of the problems it seeks to resolve. But one has also to realize that the usefulness of all the promising qualities of agroforestry lie in its effective practice. The effectiveness of its practice, on the other hand, depends on the capacity of the project to remove structures of inequality in the forest based communities.

Agroforestry and the poor

The rapid deforestation of the tropics has often been blamed on the efforts of the poor to search for fuelwood. But recent research has found that far more forests vanished because of the need of poor people, both in lowland and upland areas, for more farmland rather than for fuels. This implies that if an agroforestry project is to be effective, it should be able to provide an alternative way for the poor to get a stable income, thus making them independent of the forest land as their major income source. This further means that an agroforestry project should not concentrate purely on tree planting but it should also contain programs which guarantee the poor enough income to support their families.

Ideally an agroforestry project should be preceded by an effective land reform program, as this would allow the poor to own their own land and thus enable them to get more benefits from the project. That land reform is a prerequisite to the success of an agroforestry project can be seen in the state of West Bengal, India. Here the poor actively plant trees on their newly acquired land. A similar occurence is reported from mainland China where peasants actively participate in the government sponsored reforestation program. The problem is that not many governments in the developing countries have the political will to restructure drastically their countries' landownership patterns.

To sum up, deforestation is part of a wider problem of rural development and therefore only when an agroforestry project is designed as a rural development project, not merely a forestry project, can it be successful. To grow and use trees the poor need not only seeds but they also need land, education and money to buy fertilizer and basic implements like saws and axes.

The forestry department and the success of agroforestry projects

The success of an agroforestry project cannot be achieved without a drastic change in the nature of the relationship that presently exists between the forestry department and the members of the forest-based community. Officials from the department continue to see the latter as a potential menace to the forest, who have to be policed constantly, rather than as partners to forestry development. In Indonesia, to protect the forest the Department of Forestry maintains its own police force called the Special Forest Police. The existence of this police force has created an image of authoritarianism among the forest-based communities.

To make an agroforestry project a success the Department of Forestry must not see its function as policing members of the forest-based communities, but more as agents of development. This means that foresters working in the Department must think of themselves as human beings first and foresters second. As human beings they must be willing to share and care about things that may not even have anything to do with the forest, that is, the welfare of those fellow human beings who live in the forest-based communities.

The Perception of Peasant Land Rights in Indonesian Development: Causes and Implications

Michael R. Dove

Winrock International Institute for Agricultural Development, Islamabad

Introduction

Controversy over land characterizes almost every government program involving agriculture and/or forestry in Indonesia today. The question in each controversy is who holds rights to the land involved and what sorts of rights these are. The purpose of this paper is to examine the extent and manner in which peasant rights are recognized in these situations and, in particular, whether the peasants are recognized as holding *hak milik* "rights of ownership" or merely *hak ulayat* "rights of avail"--as the latter term is often translated now (Holleman 1981). The thesis of this paper is that the perception and treatment of peasant rights is determined largely by the interplay between the interests of political-economic elites, development officials, and the nation as a whole. Attention to the reality of this situation, as opposed to the ideal of how development programs are supposed to work, offers important lessons for the design of all programs involving peasants and land.

Background

The relevant history in this area of Indonesian law begins with the Agrarian Act of 1870, passed by the Dutch colonial government. This act guaranteed full protection of the rights of Indonesia peasants to land that they kept under constant cultivation or other use; but the so-called "virgin or waste lands"--which included all fallowed swidden lands, for example--were thenceforth designated as state dominions, to be disposed of as the state wished (Geertz 1963, Holleman 1981). The latter provision of this law initiated a century and more of conflict between the state and swidden farmers, particularly in areas designated by the state as forest preserves. Then, in 1960, the national government passed the *Undang-Undang Pokok Agraria*, "Basic Agrarian Laws," one of the explicit purposes of which was to do away with this provision of the 1870 law classifying unused lands as the dominions of the state (Brewer 1985). This was accomplished in part by according explicit recognition to *hak ulayat* or "rights of avail", which--as is explicitly noted in the "memorium" to the 1960 laws--were never recognized in law by the colonial government (*Peraturan Dasar*:31). However, in the same article in which these rights are recognized, it is added that this recognition must be in accord with national interest and national unity, and that it cannot conflict with "higher" laws and regulations (*Peraturan Dasar*:5).

Even with this added caveat, the law still reads to the effect that rights of avail must be recognized except in particular circumstances when it can be proved that this recognition would conflict with the national interest. In practice, however, the burden of proof is reversed. All development officials know the wording of this article by heart, and they take it to mean--and they in fact employ it as meaning--that whenever and wherever rights of avail conflict with their projects, these rights can automatically be ignored or overridden. There is no awareness whatsoever that development planners need to *prove* that national interest will otherwise suffer before this article of the law can be invoked. I am not aware of a single study, analysis, or even verbal statement to suggest that any official has ever tried to meet the burden of proof in such a case. The burden should not be on the holders of the rights of avail, but it is, and it is clearly a weighty burden. I am not aware either of a single instance in which holders of rights of avail have been able to officially contest the presumption of national interest in such a case. This failure to either prove or contest these claims of national interest obviously raises the possiblity that such claims are sometimes used to override rights of avail for purposes other than the national interest or even contrary to the national interest.

Many development planners in Indonesia clearly do not understand what rights of avail are. In particular, they commonly err in perceiving these rights as not merely different from but opposed to *hak milik* "rights of ownership." This is reflected in the Basic Agrarian Laws, which treat the two types in completely separate articles, as well as in discussions of the subject by development officials, which make clear that a given area or people is assumed to be subject to rights of avail *or* rights of ownership but certainly not *both*. In fact, the two types of rights have occurred congruently, as part of a single land tenure system, at many times and in most parts of Indonesia. (This congruence was also a source of confusion for Dutch colonial administrators: see Holleman 1981). Among the contemporary tribal

Kantu' of West Kalimantan, for example, individual households hold rights of ownership or "primary" rights to fallowed swidden land, while the long- house holds rights of avail or "residual" rights to the same land (Dove 1982). The household rights cover use and devolution, while the longhouse rights cover land left behind by a household that has moved out of the longhouse. In the vast majority of instances in which development planners identify *hak ulayat* "rights of avail" in contemporary Indonesia, these are similarly only *one* aspect of the local system of tenure to the land in question, and not the most important aspect at that.

There is also a great deal of peasant land in Indonesia to which rights of avail do *not* apply, but *only* rights of ownership. Among the aforementioned Kantu' of West Kalimantan, for example, fallowed swidden land in which rubber trees (*Hevea brasiliensis*) have been planted is thereby reclassified as *tanah mati* "dead land," meaning that any and all residual rights to that land on the part of the longhouse have ceased to exist: it is completely and forever under the control of its owner(s). Even in cases such as this, however, in complete defiance of all that is known or can be observed regarding tribal or peasant systems of tenure to land and plants, there are development officials who classify such land as subject only to *hak ulayat* "rights of avail." In the most recent case known to me, this was used to justify the appropriation and clearing of 2500 locally owned rubber trees in West Kalimantan by a state plantation (see *Sinar Harapan* 1985). This was not an isolated case. Wherever land is needed for some development project, whether government or private, and especially in Indonesia's outer islands, land whose ownership has not been registered with the proper authorities--which means virtually all of the peasant lands in these areas--is classified as falling under only *hak ulayat* "rights of avail."

Analysis

Assuming that government policy--both the stated policy as well as the actual policy--is somewhat purposeful, there are two possible explanations for the treatment of peasant land rights that has just been described. One explanation is self-interest: development officials generally prosper when the projects in which they are involved prosper, and it is far easier to establish a transmigration village, a commercial logging concession, or an export crop estate in the outer islands, if local claims to the land involved--and there are always such claims in the outer islands, popular myths about their underpopulation notwithstanding--can be regarded as based on *hak ulayat* "rights of avail," as opposed to *hak milik* "rights of ownership." The former can be routinely overridden, as noted earlier, but the latter cannot. As stated in the Basic Agrarian Laws, *Hak milik adalah hak turun-temurun, terkuat dan terpenuh yang dapat dipunyai orang atas tanah* ("Rights of ownership are hereditary rights, the strongest and fullest that can be held by a person to land") (*Peraturan Dasar*:10). It is generally recognized that when rights of ownership are involved, even the government cannot appropriate land without first paying compensation. Payment of compensation would make all development projects more expensive, and it would make many completely impractical. Hence the developers' vested interest in perceiving rights of avail as opposed to rights of ownership.

While self-interest is one determinant of this perception of peasant land rights, it is not the sole determinant. There is a second determinant or explanation that involves, paradoxically, the public interest. Many development officials maintain that the problem in paying compensation for rights of ownership in government projects--and by implication, the problem with recognizing such rights in the first place--is that local *orang besar* ("big men") will seek to take advantage of the situation. And indeed, there are increasing numbers of incidents in which local officials and in particular ex-officials (who still have some of the power of office but are no longer subject to its code of conduct) plant surveying poles, clear the forest, and erect fences on land that has been selected for a government project. These land improvements are undertaken in the hope that the government will then view this land as privately owned (*hak milik*) by the officials and ex-officials in question, and will be obliged to pay them compensation for it. The tactic of improving land in order to influence the government's perception of its ownership status is becoming increasingly common. For example, Banjarese swidden farmers in South Kalimantan plant fruit trees on their fallowed swidden land largely in order to force local government officials to classify that land as subject to *hak milik* "ownership rights," or minimally *hak ulayat* "rights of avail," as opposed to the usual classification of all forested land as *tanah negara*, "state land" (Dove 1983). The difference between these two examples, of course, is that the latter represents an attempt to force the government to recognize the rights of the true owner of the land, while the former represents an attempt to appropriate land that belongs either to the state or, more likely, to local peasants.

Such attempts to deceive the government into paying land compensation to nonowners typify what transpires in Indonesia in any situation involving the development or in effect, devolution of land. Consider, for example, the government's land registration program in Java. According to the intent of this program, it is simplicity and justice itself: on specific dates made public well beforehand, the inhabitants of each village come together and make their respective claims before the land registration officials, thereby protecting forevermore their rights to their own land. In practice, in at least a significant minority of cases, the dates of the program are not announced except to local elites, who alone then meet with the land registration officials and make false claims to as much of the local land as possible. In these cases, then, the program to register land in effect gives powerful elites an opportunity that they would not otherwise have to steal it. The general principle that is operating here, as well as in the previous example of improving land to support false claims for compensation, is that whenever the government is involved in the devolution (and/or development) of land, an opportunity is created for economic and political elites to misappropriate it. They have the resources and --where the land is in fact not theirs--the motive to do this, whereas the local peasant owners have neither.

To return to the original question as to why development officials tend to classify land targeted for development as subject to *hak ulayat* "rights of avail" as opposed to *hak milik* "rights of ownership": this is in part a response to the problem just described. That is, by classifying the land rights involved as rights of avail rather than ownership, the central government can simply appropriate the land in question, thereby not raising the issue of compensation nor, consequently, creating any opportunities for misappropriation by local elites. To the extent that central government officials are successful in this tactic, inequities in the local distribution of land and utilization of government resources are, minimally, kept from growing worse. So the government's misperception of peasant land rights in these cases is, on balance, in the public interest (admittedly it is likely to be in the government officials' own interests as well): land is appropriated for national development, and although the deserving peasants receive no compensation, neither do the undeserving elites. It bears noting, however, that the poorest and least influential Indonesians thereby wind up paying the brunt of the cost of controlling the rapacity of the richest and most influential.

Summary and conclusions

At the beginning of this paper, I discussed the treatment of *hak ulayat* "rights of avail" under the colonial Agrarian Act of 1870 and the post-independence Basic Agrarian Laws of 1960. I noted that in both colonial as well as contemporary times, central Indonesian governments have tended to see rights of avail and *hak milik* "rights of ownership" as distinct and separate systems of land tenure, as opposed to complementary aspects of a single system; and, in consequence of this error, they have tended to interpret most rights of ownership as rights of avail. I went on to explain this misperception, at least on the part of contemporary officials, in terms of self-interest in improving the economics of development projects, and public interest in reducing opportunities for land grabs by political and economic elites that otherwise accompany most government devolution and/or development of land.

From this analysis, we can draw one major conclusion of relevance for development planning in agroforestry: First, any government project involving the clearing, registration, redivision, resettlement (etc.) of land is potentially disastrous for the local peasant land owners. Put in more general terms, in any sort of government sponsored program of land development, those who have no special access to information and influence will lose out to those who do have such access --regardless of the relative justice of their cases. One immediate implication of this conclusion is that agroforestry programs, at least in Indonesia, should focus at the level of the individual farmer on extant holdings. More ambitious programs, involving relocation of peoples and redistribution of land, should be avoided like the plagues that they are for the peasants affected. A second implication, therefore, for both the government and international aid agencies as well, is that greater expenditures do not guarantee better development. Indeed, experience in Indonesia shows that it is difficult to spend large amounts of money without being, in effect, destructive. A third implication, of special relevance for the international aid agencies, is that appearance and reality in the developing world are often --indeed, perhaps most often--not the same thing. An agency like my own country's USAID jumped on the land registration bandwagon in Indonesia, secure in the belief that it was helping "the poorest of the poor," as its current mandate demands, whereas some of its assistance may in fact have gone to help "the richest of the rich." Good intentions, therefore, when they focus on appearance and ignore reality, can be worse than no intentions at all. Good intentions must be informed by a sophisticated knowledge of, in particular, the political economics of developing countries like Indonesia--a sophistication that in many cases is still lacking.

References

Brewer, J.D. 1985. Traditional land-use and government policy in Bima, East Sumbawa. In M.R. Dove (ed). *Peranan Kebudayaan Traditional Indonesia Dalam Modernisasi* (The Role of Traditional Indonesian Culture in Modernization). Obor Foundation. Jakarta.

Dove, M.R. 1982. The myth of the communal longhouse in rural development: The Kantu' of Kalimantan. In Colin MacAndrews and L.S. Chin (eds). *Too Rapid Development.* University Press. Athens, Ohio.

Dove, M.R. 1983. Theories of swidden agriculture and the political economy of ignorance. *Agroforestry Systems* 1:85-99.

Geertz, C. 1963. *Agricultural Involution: The Process of Ecological Change in Indonesia.* University of California Press. Berkeley.

Holleman, J.F. (ed). 1981. *Van Vollenhoven on Indonesian Adat Law.* Koninklijik Instituut Voor Taal-, Land- En Volkenkunde Translation Series No. 20. Martinus Nijhoff. The Hague.

Peraturan Dasar Pokok Agraria (The Basis of the Regulations Regarding Agrarian Principles). n.d. C.V. Patria. Bandung.

Sinar Harapan. 1985 (20 April). 2,487 pohon karet penduduk tergusur proyek PIR khusus Sintang PTP XIII (2,487 of the local inhabitants rubber trees destroyed by the nuclear estate project in Sintang State Plantation XIII).

Agroforestry and Land Tenure Issues in the Philippines

Percy E. Sajise

Program on Environmental Science and Management, University of the Philippines at Los Banos

A basic premise concerning the relationship of agroforestry (or any other form of resource use) and land tenure is that a stable, sustainable, productive and socially acceptable form of agroforestry will evolve under a favorable land tenurial system. Is this premise correct? If it is, then understanding this relationship will be important in identifying intervention strategies to promote agroforestry programs.

Some lessons from traditional systems in the Philippines

In the Philippines, there are generally two groups of cultural communities that occupy the uplands: 1) those who were able to maintain their customary laws and to control and maintain their territorial boundaries, and 2) those who have been eased out of their original territorial boundaries by the lowland Christian migrants.

The Bontocs

The Bontocs of the Cordillera, Northern Luzon are a case in point. They have customary rules and ownership forms that govern the use of "fatangan" (forest lands), "punching" (pasture land), "uma" (swidden land) and "payew" (irrigated land).

Land tenure is based on the establishment of rights prior to exploitation, on the defense of territory prior to encroachments, and on the development of land tenure rules which govern the rights to communal land, corporate land and individually-owned land. It is tied to economic considerations, social customs and political interests. The Bontoc case, exemplifying traditional tenure patterns and resource use, has shown that the community can utilize and protect the forest at the same time. They practice indigenous agroforestry and sustainable rice paddy cultivation as individuals, kin groups and as a community. Yet, in the context of existing national laws, the Bontocs cannot own their land because its slope is more than 18 percent and therefore is in the public domain.

Issue: How can customary law be accommodated in the existing national system by using suitable indigenous agroforestry and land use system as a basis?

Note: Integration of customary law into the existing legal system of land tenure has already been done in Indonesia.

The Palawanons

The case of the Palawanons of Palawan Province, southwestern Philippines, is somewhat different because they represent a cultural community that has been pushed by lowland Christian migrants into the mountainous areas. Being relatively inaccessible, they have been free for some time from the further encroachment of lowlanders. They have developed indigenous forms of agroforestry in which local fruits and sugarcane, sweet potato and other vegetables are grown in sequence or intercropped. These practices have diffused because of the belief that when somebody in the family dies, they have to transfer to another place. Thus, there are pockets of agroforestry. However, the Palawanons are conscious that they do not own the land.

The present program of the government known as Integrated Social Forestry (ISF), grants them a lease of 25 years, renewable for another 25 years.

Issues: What form of stewardship contract is most appropriate for this kind of group? Should it be an individual or communal lease? What are the necessary guidelines for each form of leasehold or stewardship contract?

The aforementioned two issues presently confront more than 5.5 million "tribal" Filipinos in the uplands where agroforestry is the most appropriate form of land use and is promoted by the national government.

Some lessons from non-traditional systems

Christian settlers have also used agroforestry techniques despite the lack of secure tenure.

The case of sustainable agroforestry practices even if the community does not own the land

There are several cases that can be cited in the Philippines where sustainable forms of agroforestry have evolved in spite of the fact that members of the community are aware that they do not own the land.

There is the case of the *Leucaena*-based fallow system in Cebu, Central Visayas region in the Philippines, where corn, tobacco and sweet potato are planted between sticks of Leucaena known as "babag" laid across the mountain slopes. This technology has been practiced by the entire community in the area for more than 90 years. They are willing to stand together as a group to claim their rights, believing that they have come up with the right form of land use for their sloping land. A similar case is provided by the community at Masalukot in Quezon (a province in southern Luzon, Philippines), where an indigenous agroforestry scheme evolved among migrants from the neighboring province of Batangas.

Since the community stands as one, the scheme evolved in spite of the fact that the area is declared a national reserve. They reason that they are not doing harm to the environment with the agroforestry scheme they have developed and are still using.

Issues: What are the necessary ingredients that will promote agroforestry even if communities and individuals do not legally own the land? Is legal ownership necessary or is it more important to have community solidarity and a strong psychological belief that they have right to these lands?

The case of land ownership hindering the development of agroforestry

There are cases in the Philippines, where agroforestry as a land use system is not adopted because of land tenure issues.

In an upland development project in Buhi, Camarines Sur, southern Luzon, the landowner will not permit the planting of perennial crops in a sweet potato-based cropping system because he is apprehensive that the tenants will lay claim to the land.

There is also a case of a coconut hacienda in Jaro, Leyte Province in eastern Visayas where the landlord will not allow tenants or caretakers to have more than two pigs because he fears that the tenants will be using the harvested coconuts as feed instead of making copra with it.

Issues: What form of crop/animal sharing is best for lowland tenants in order to support the development of agroforestry schemes in tenanted areas?

The case of the multiple land use pattern and the role of agroforestry

In many parts of the Philippines, even in upland areas, farmers are actually engaged in several forms of resource use. An individual may cultivate a lowland rice field and an upland field, use an upland field and fish at the same time, or engage in various other combinations of multiple resource use. This means that a farmer, with limited labor supply and resources, may own land in one place and be a tenant in another place at the same time. It is interesting to note that one tenant farmer took advantage of an upland agroforestry project, consisting of strip cropping of *Leucaena* and upland rice, by planting the new upland rice variety in the lowlands where he owned a rice paddy.

Issues: How do you develop agroforestry as a complementary resource use system for a farm family household with multiple resource use patterns and multiple land tenurial statuses?

References

Buhi-Lalo Upland Development Project. 1983. Annual report. Mimeo.

Farming Systems Development Program -- Eastern Visayas, 1983. Annual report. Mimeo.

Lynch, O.J., Jr. Forthcoming. Philippines law and upland tenure. In S. Fujisaka, P. Sajise and R. del Castillo (eds). *Upland Development in the Philippines*.

Palawan Integrated Area Development Program. 1982. Annual report. Mimeo.

Prill-Brett, J. 1985. Bontoc land resource and management. Paper presented to the 2nd SUAN-EAPI Regional Research Symposium. Baguio City, Philippines.

The Land Tenure System in the Philippine Uplands: Implications for Agroforestry

Luzviminda B. Cornista

Agrarian Reform Institute, University of the Philippines at Los Banos

The purpose of this paper is to describe briefly the land tenure system in the Philippine uplands as it relates to agroforestry and to pinpoint relevant research and development issues with regard to tenure and agroforestry.

The structure of landownership in the uplands

In the Philippine uplands, lands are classified either as privately owned or part of the public domain. Of the total 30 million hectares comprising the country's land mass, 18.69 million hectares (or 62 percent) are considered public lands of which 11.07 million hectares are timberlands and 5.5 million hectares are still unclassified (NEDA 1984). There are no available data on the total upland area under private ownership. A tentative estimate is that 10 percent of the country's moderately to steeply sloping land is privately owned.

A problematic issue in landownership concerns the case of ancestral lands inhabited by tribal minorities. Their claim to ownership of their ancestral lands is not substantiated with land titles and other legal documents. Thus, the Revised Forestry Code has declared ancestral lands as part of the public domain. A Supreme Court decision in the Carino case of 1909, however, reiterated that lands held by native custom and long association under a claim of private ownership are presumed not to have been public lands. This ruling has not been enforced. A consequent effect of this situation is the displacement of a large number of tribal minorities from their abodes and livelihood. Among the causes of displacement are the encroachment on ancestral lands by lowland farmers, forest concessions and illegal logging, government projects such as national parks and resettlement, and expansion of agribusinesses (Lynch 1984).

Land tenure patterns in the uplands

There are several distinguishing characteristics of the land tenure system in the uplands which can have a direct bearing on agroforestry development.

First, in both public and private lands, the presence of share tenancy is apparent. Share tenancy in public lands usually exists in areas where there are non-cultivating claimants to the land. Claimants are those who have "ownership" claims on some public lands made through their continued payment of land tax to the government. These non-cultivating claimants are generally absentees and act as virtual landowners.

Second, the sharing arrangements in tenanted areas vary by region and by crop. For instance, coconut in the southern Tagalog region is shared on a 6/7:1/7 basis in favor of the landowner; in the Bicol and Visayan region, the sharing is usually 1/3:2/3 in favor of the tenant. In the former, the net product is shared; in the latter, the gross produce is divided with all harvesting and copra processing costs borne by the tenants. Perennial intercrops are usually owned by the landowners. Coconut tenants get their share of the perennial intercrops if they take care of the trees. It is possible that another person, not necessarily the coconut tenant, gets the share from taking care of the perennial intercrops. In a village in Cebu, the sharing, regardless of the kind of crops raised, is 2/3:1/3 in favor of the tenant with all expenses shouldered by the tenant. In another village in Camarines Sur, rice, corn and bananas are shared on a 3/4:1/4 basis while abaca is on a 1/2:1/2 arrangement.

Third, in general, upland farmers are essentially subsistence farmers. This is manifested in the farming system they have adopted. In their *kaingin* (slash-and-burn) farms, they first grow rice, corn, root crops and vegetables for subsistence. Perennial trees such as coconut, cacao, coffee and other trees are grown only if the soil can no longer sustain the subsistence crops. Even where fixed agriculture is practiced, upland farmers still maintain a substantial portion of their farms for subsistence crops.

Fourth, the advent of commercialization has made farmers produce for the market. For the small upland farmers, the more convenient crops for market are primarily the short term ones (e.g. vegetables) which usually require less capitalization but provide immediate income returns.

Fifth, the Integrated Social Forestry (ISF) program is a recent attempt to provide forest-occupants access to public (and/or ancestral) lands, guarantee security of tenure on the said land, while at the same time arresting ecological degradation through agroforestry development. There is a stewardship agreement between the government and the forest occupants allowing the latter continued tenure on the land for 25 years renewable for another 25 years. The stewardship contract can be granted to individuals, communities (usually tribal minorities) and, in a more recent scheme, to associations.

Lastly, the foremost aspiration of uplanders, whether those found in private, public and ancestral lands, is to gain security of tenure on the land they occupy. However, for those whose tenure is already secured, for instance, those who have been granted titles or stewardship contracts under the ISF program, their overwhelming desire is to uplift their socioeconomic condition. They perceive increasing productivity as the proper strategy to attain this goal.

Research and development implications

From the above several implications can be gleaned for research and development in tenure and agroforestry.

1) There is a need to recognize the existence of an indigenous system of property rights among tribal minorities which may not necessarily conform to the national legal system. An understanding of the interplay of these varied systems is important to agroforestry development in tribal communities. Equally necessary is the resolution of the issue of ownership of ancestral lands to arrest the displacement of ethnic minorities. An essential research area along this line, as pointed out by Lynch (1984), is the interface between indigenous and national law.

2) The issue of share tenancy in the uplands and its effects on the adoption of agroforestry technologies should be further investigated. In areas where there is no assurance of sharing of the produce or where the sharing system is perceived to be inequitable, tenancy can be a disincentive to the acceptance of agroforestry technologies. An emerging phenomenon which is an indirect offshoot of the implementation of the agrarian reform program is the fear of owners (and even claimants) that by allowing their tenants to plant perennial trees, these tenants would have a more permanent claim on the land.

There is another form of "tenancy" arrangement that should be studied carefully to find out its implications for the implementation of an agroforestry program. It is observed that in some ISF areas, participants lease out portions of the land grants to other farmers with miniscule farms or to the landless. These farmers are usually relatives or friends of the participants. In a sense, this situation can be viewed as a survival strategy among poor upland farmers and a means of sharing poverty.

3) The subsistence orientation of upland farmers should be a consideration in framing any agroforestry program. The transformation of their economy into a market-oriented one through the introduction of a farming system primarily geared towards the market should be taken with great caution. Completely exposing upland farmers to the variability of market forces might result, in the long run, in further worsening of their socioeconomic condition.

4) Aside from tenure considerations, the adoption of agroforestry technologies by upland farmers is also premised on some economic considerations. For instance, the planting of *ipil-ipil* (Leucaena) trees is already accepted by upland farmers as ecologically sound as well as economically viable in the long run. However, participants in such tree planting programs will still tend to allocate a substantial portion of their land to short term crops for their immediate cash needs and to subsistence crops for their household consumption. The challenge to agroforestry development is, therefore, to come up with technologies that are not only ecologically sound but also economically viable and socially acceptable as well.

5) Increasing productivity in the upland areas is an issue that will have to be dealt with in the immediate future by developmentalists. Ultimately, sustaining the gains in agroforestry will be largely premised on whether the socioeconomic condition of the upland farmers has been favorably affected by such a program. This, in turn, is primarily determined, aside from the equity factor, by economic productivity considerations. Concomitantly, there is the need to develop functional support institutions to productivity, particularly credit and marketing.

6) It is necessary to study more closely the land tenure improvement schemes being articulated by the ISF program. It would be interesting to find out how the different types of stewardship land grants (to individuals, communities, and associations) could affect agroforestry development.

References

Castro, P. C. (ed). 1984. *Uplands and Uplanders: In Search of New Perspectives*. BFD Upland Development Program. Quezon City, Philippines.

College of Law, University of the Philippines. 1984. *Mga Karapatang Legal sa Lupand Ninuno*. Indigenous Law Project. Quezon City, Philippines.

Lynch, Owen J., Jr. 1984. Withered roots and landgrabbers: a survey of research on upland tenure and displacement. In P.C. Castro (ed). *Uplands and Uplanders: In Search of Perspective*. BFD Upland Development Program. Quezon City, Philippines.

National Economic and Development Authority (NEDA). 1984. *Philippine Statistical Yearbook*. Quezon City, Philippines.

Agroforestry and Swidden Cultivators in Latin America

Rafael Baraona

Cuernavaca, Morelos, Mexico

Introduction

As I understand it agroforestry is a relatively new and integrative scientific discipline, struggling for recognition and application to real problems. Agroforestry has very old antecedents and agroforestry practioners are considered by many as mere relicts bound for extinction. I refer to tropical swidden cultivators. The scientific agroforesters and their native counterparts face an uphill battle for survival or for recognition. It is no small matter to postulate that field and forest should survive together and, better yet, be integrated into productive systems, when progress or modernity have decided that field and forest should be torn asunder.

To me, peasants and particularly swidden cultivators appear as the natural target beneficiaries of agroforestry. The latter share with the scientific agroforester a community of aims and of subject matter.

These days the paths that lead to technological adoption are very much like a maze and we normally detect the exit. The technical messages come from everywhere, the bombardment of advice and the sales pitch is unceasing--eventually reaching even peasants. In this surfeit of information the message of agroforestry is rarely heard. So we presume that for agroforestry it doesn't make sense to compete in this game. No doubt here in ICRAF other means of communication are being thought out and tried.

In this paper I am suggesting that certain situations should be taken into account in order for scientific agroforestry to reach peasants effectively. At least in Latin America, conventional agronomy and allied disciplines have mostly proven unable to foster the adoption of technical advances in agriculture or agroforestry among peasants; that is, advances that are really viable and suitable. At times, we have known the what but not how, and at others, knowing how, we didn't exactly know what. Agroforestry should learn from these experiences. But we also know that major forces were at play and there is no point in attributing blame to agronomists or to peasants. We are quite aware that available technology will be eventually adopted, if perceived by peasants as suitable.

One plausible way for scientific agroforestry to be diffused among swidden peasants is to work towards a built-in sort of strategy, one that is linked to its development as a scientific discipline. We have at our disposal many contributions that could be used and integrated in designing such strategy.

This workshop deals with the role of land tenure in agroforestry and this very choice of theme shows that the broad concern is with the future of agroforestry. The proper consideration of the institutional framework of access to and the use of land and trees and their products, and the arrangements between persons or groups in land use operations is paramount. They are properly identified and assessed in the background papers and we will be learning much worth sharing from the position papers. This position paper should be seen as a complement to both concerns: land tenure and the future of agroforestry, as a discipline that should find ways to become accessible to peasants.

The crises in swidden cultivation

Swidden cultivation, under functionally appropriate conditions, can offer a number of advantages to the tropical peasants of today, just as it did in the past to tribal cultivators. Firstly, and significantly, it does not require heavy outlays in farming equipment. In Mexico, cash needs are of some impact and they often limit, where wages are paid, the acreage of land that can be cleared. When other sources of income are available, shoulder pumps and herbicides are employed. Swiddens are considered energy efficient and there is no point in devoting energy or space to traction animals. Imports or subsidies of fossil fuels or fertilizers are simply not needed, as long as the conditions for viability in swidden persist.

In sum, shifting cultivation is in all likelihood, because it is so autonomous, *peasant agriculture* par excellence. It allows cultivators to carry on with production using simple tools, applying what they already know, and with a minimum of added risks. This is when things are going well. But this condition is rarely found today.

What is happening to swidden cultivators all over Latin America is not really different, in essence, from the fate of other peasant agriculturalists in the world, whether they practice permanent field agriculture or various systems of fallow cultivation. Social scientists have described the symptoms and the causal processes: proletarianization, depeasantization, decomposition of the peasant unit, etc., etc. all this under the impulses of modernization, urbanization, etc. Some scholars have predicted the imminent demise of peasantries but others do not quite agree and the polemics could fill several volumes. Unfortunately, to counterbalance this abundance, we do not have enough precise information on how peasants are facing the onslaught of modernity; and so we do not know how the survivors manage to survive.

If we examine very briefly some aspects of what has been called the "crisis of *roza-tumba-guema* agriculture," which in some degree are likely to be found all over the American tropics, we might be able to highlight where we think the aims of scientific agroforestry and the needs of swidden cultivators should meet or coincide. This might provide a starting point for what needs to be done provided we see it in the context of the whole problem of peasant agriculture and survival.

With some limitations, extreme or crisis situations can be used to make more vivid a premise or assumption. When the extreme effects of territorial restrictions on a population of swidden cultivators are felt, as in some areas of the flood plains of Tabasco, new qualitative conditions emerge. In facing them the swidden cultivator becomes bewildered, puzzled or just feels helpless. He is simply unable to cope with the new conditions and pays dearly for the passing of a very autonomous system; autonomy which made it the more vulnerable.

Gone are the complex interactions and diversity of environmental materials that gave relative stability to his swiddens. I have heard peasants giving precise recognition to the fact that they no longer have the tools, the skills or the knowledge to face or understand this new nature: generalized plagues, unheard of before, spreading from the encroaching grasslands to their milpas; changes in soil structure; weedy secondary growth; persistent grasses with unwieldy stolons, etc.

The body of knowledge

The cultivators had depended on a biological diversity that no longer could be recovered. Information can be considered as the key net loss resulting from the vanished forest--their whole cognitive system was a sort of intense and complex recognition of the main features and processes contained in a diversity no longer present. Gone are the references, clues and indicators implanted in their corpus of knowledge, which in turn had provided answers to the praxis which also had the role of keeping the corpus alive and valid.

For these swidden cultivators their corpus no longer provided the solutions that their customary praxis had obtained. Or worse yet, they found themselves without questions to ask, and for peasants this is a most dangerous and vulnerable situation. This is not likely to happen in such a "no exit" way to other, non-swidden peasants. Their cognitive systems do not become obsolete in those terms because, among other reasons, their environment materials never embodied comparable information, and their cognitive systems as such shared many perceptions with non-peasant systems and were amenable to a greater variety of uses.

On the other hand, to become depeasantized ex-producers is a common enough happening that generates a whole range of intermediate situations and new versions of survival alternatives. The problem is that among swidden cultivators not only those that get poorer through external pressure on intensified internal differentiation become ex-producers. It tends to cover the whole range of peasant stratification. Here the moot question is the timing for scientific agroforestry support and of course content and an *ad hoc* strategy.

In this respect the situation of Maya catecan swidden cultivators poses some different clues. Their nonconventional scientific system rests on different tenets. The main perceptions of the deltaic farmers of Tabasco deal with plant indicators for site selection and the adjustment to periodic flooding where barely perceptible relief differences take place within large scale landscapes. The Maya are possessors of a particularly rich ethnobotany and some of them can still be considered very knowledgeable high forest managers. But most Maya live in ancient permanent villages with assigned territories covered with mostly degraded secondary growth. Today they are masters of Hub'che (secondary growth) cropping. Their milpas are not particularly rich in diversity--nothing like those of the Lacandon Maya. Their main perception concentrates directly on the soil as an autonomous phenomenon and have probably one of richest and least understood ethnopedologic and ethnoedaphic knowledge systems in the world, based on distinguishing soil micro differences even in very small areas. This probably explains why they tend to maintain several separate crop-specific clearings. This could be read as a different revolutionary tendency suggesting different contents in technical support. Yields are universally decreasing but there is no agreement as to the causes.

If the *roza-tumba-guema* crisis is generalized and no viable technical support is forthcoming for productive alternatives, ex-swidden peasants have to take whatever options are offered by the very same forces responsible for the crisis (aggressive ranching and the so called "humburger connection"): sale of labor force, involvement without participation in unsound and ambitious large-scale projects, or very unprofitable share arrangements with ranchers.

But the main loss, as already suggested, was the obsolescence of their functional system of corpus-praxis interaction. The impact of external processes deprived them of their own cognitive or scientific system but no viable replacement was proposed. Peasants can manage to survive with minimal capital but they cannot exist without their science.

So if we accept that some kind of science to be depended upon lies at the core of peasant identity, at least two topics should be considered if we intend to work with their agriculture or agroforestry in support of their livelihood: 1) the points of coincidence and of disparity between their science and our conventional science, and 2) the position of their cognitive or scientific system in their strategy for survival.

At this point I borrow freely from ethnoscientific concepts and categories but what I infer from them is my own responsibility. The native scientific system rests on a body of relatively shared knowledge with the appropriate name of corpus. Relatively shared does not imply, for me at least, that some peasants know a lot less than others, but that there is no complete agreement among corpus possessors about meanings and alternatives. The corpus has a real but implicit existence; its locus is the memory of its possessors. If studied from the outside it has to be reconstructed and, for some purposes, translated and made intelligible to conventional science.

I suspect that unaltered swidden systems have the rare property, in common with conventional science, of being mostly contained in one consistent paradigm. In Latin America other cognitive systems are in varying degree, syncretic--combining Spanish--Mediterranean and New World perceptions and traits. Although some internal ordering can be discerned, these other cognitive systems serve as assortments of internally linked clusters of solutions to choose from. These systems are less vulnerable than swidden systems, as they have more areas where perception can be accrued, stored or preserved and, of course, forgotten. So even if less internally consistent they tend to be more functional to peasant survival.

Keeping our purpose in mind, change within the cognitive system operates from praxis to corpus. I would add further that we are mostly interested in praxis as our starting point for inquiry and that a key problem in this kind of research is to identify critical or revealing sets or sequences of praxis that let us infer about corpus limitations or potentials.

The main task of the peasant is to make viable through praxis his nonconventional scientific system, and so this system, is constantly being tested for its relevance in survival. The main purpose of conventional science is to make more science, according to established or emergent paradigms, and so change can best be understood as the history of ideas. In the peasants' case, the dynamics of nonconventional systems are probably best understood in relation to ever-changing objectives and material conditions (including options and constraints) and not as an autonomous process of paradigmatic change.

From this assumption we can locate another source of discrepancy --the rhythm of obsolescence of conventional technical advice hardly ever coincides with its possible absorption and rejection in peasant agriculture. Peasants know from experience that they cannot depend on being up to date in technology but must combine, in different formulas, the old and the new (both seen as partial solutions). There is nothing more risky for a peasant than giving up his main resource: his memory. He knows that sooner or later he will have to probe into the corpus for a solution to a problem not frequently faced.

What this means for tenure

In Mexico most peasant landgroups occupy territories assigned to them in the long process of land reform or are regulated by special federal legislation. And so it is with most swidden cultivators. Commercial farms and ranches survive and probably increase under different guises and in a fashion adjusted to the times are likely to be just as dominant and influential as haciendas were in the past.

Here as in other countries of Latin America we have a rich field of observation for contradictions, discrepancies, adjustments and trade-offs between land tenure institutions, mostly the product of urban goodwill and imagination, and peasant realities. Social planners and others have never given up the hope that peasants, once they become beneficiaries of land redistribution, will behave better or at least a little differently. This last hope is usually fulfilled: peasants learn to adjust and to take advantage of new links with government institutions.

The *ejido* really acts both as source of security for land possession and as a container for peasant interactions. This interaction, (factionalism or internal differentiation notwithstanding) is mostly *interpares* and they tend to agree on how to face the outside world. Although some *ejidatarios* profit more than others from these contacts.

If it fits their survival strategy *ejidatarios* are willing to rent land to outsiders. They also get together in production when seen by some or by all as feasable or convenient. Most are willing to go along with the fiction of the *ejido colectivo* but they carry on with production as usual, on a family basis or by ad hoc arrangements.

Government regulations that go against the grain of peasant behavior are simply unheeded and cannot be enforced. Peasants can be affected and even done away with by present society. But as a social group they antecede it and cannot be tamed or be "created" like industrial workers.

All these considerations should not obscure the fact that shifting cultivators, though constricted in space, are secure in the possession of their territory. It cannot be taken away from them without conflict or violence. This situation should be treated as a given fact when considering policy. Outside economic forces intervene inside the *ejido* and, depending on perceptions of alternatives, they might be considered as options by *ejidatarios*. Other complexities also play a role, but the safest guess is that those who yield to these temptations do so because they have problems in carrying on production on their own. That is, in their survival strategy the core of the corpus-praxis interplay, or their cognitive system, is not playing a relevant role in the process of becoming or remaining a peasant. This is a most important clue to those who attempt to do something in support of peasant production.

Some land tenure and related problems need sometimes a very fine focus to be detected, as they are nearly invisible. According to common belief a Mayan peasant can clear and plant his *milpa* anywhere within his *ejido* and this is what he normally answers when asked about the rules: there are none. Or at least, he is no more willing to talk about them than he is to describe his table manners. These rules are surely known by every Maya and in a way belong to the realm of "good manners." They aren't worthwhile explaining and so it becomes a very upsetting and serious problem when some *ejidatarios* are willing or forced to disregard them.

Conclusions

We get close now to the conclusion of this rambling discourse in the search for clues for how to establish scientific agroforestry as a support discipline for peasants. We have suggested that peasant behavior is informed by corpus-praxis interaction. We might start research by detecting critical questions about praxis and then go on to examine the suitability of what we have inferred about the corpus. So far these are problems of technological alternatives and choices. But are decisions about technology taken in the isolated realm of technical-scientific thinking? We know that this is not usually so. They are taken in an area that, for lack of a better term, we will refer as common sense thinking or behavior. These stem from the peasant estimation or perception of his assets and shortcomings, and of his chances, risks, constraints and expectations. This is a lot to lump together under peasant common sense perceptions or motivations. It includes economic logic, assessment of needs, strategies for deployment of family and other labor, etc. All these topics are well covered by several disciplines and it would not be a great task to group and order what I am lumping together into this potpourri of common sense. I also suspect that in some instances the distinction between common sense and technical scientific judgement is not easy or worthwhile to make.

It is in the nature of a native cognitive system applied to activities such as agriculture or agroforestry that it cannot be understood as a living system without reference to what we call common sense and probably it is unavoidable that this requirement has the double function of making the task more difficult but more fruitful at the same time.

Our reluctance to put common sense behavior first in the itinerary of research is because we tend to take to the field too heavy a load of assumptions, and if we are not careful we might end up by confirming too many not surprising correlations.

We are very much in need of new frontiers in peasant studies and we should be tolerant when a unitary approach is put forward. We might come out with something more than just fragments of our subject. If we are able to articulate technical-scientific and common sense behavior we might come out with a peasant we know little about-or at least someone whose use and needs of scientific-technological knowledge have been identified and partly diagnosed with his help. Once this knowledge has been apprehended we might use it to query the social and production structure in which the peasant finds himself. From there we could proceed backwards to the technical-scientific realm asking ourselves new questions.

Would individuals with similar technical-scientific behavior exhibit comparable locations in the socioeconomic structure? Both yes and no answers should suggest further inquiries until the position of technical-scientific behavior is properly understood within the total strategy for survival. Eventually, I am sure we would have the clues as to how best to intervene and to whom such interventions can be directed.

In our research in Chile we have been trying, in a very tentative and experimental fashion, to apprehend peasant behavior as the sequence or articulation of the scientific--technical and so called common sense phases of thinking and to use this perspective to analyze peasant social structure. This could be formulated and done in many different ways, but we feel that what we are looking for is a reasonable approximation to the peasant paradigm in real life. That is, his path to become a social subject.

To end with a repetition, when peasants still include the technical scientific dimension in their strategies for survival many ways are still feasible to give support to their production.

We know that there are many large and powerful societal processes which are affecting peasants, mostly in a negative way. This we either cannot change, or is not up to us to alter, but the peasant's position will be stronger and his survival strategy made more effective if we can make available to him the right kind of conventional science to enrich his own science. I feel this is the very minimum we can attempt.

Land Tenure and Agroforestry in Central America: The Case of Honduras

Sergio Sepulveda

CATIE-Turrialba, Costa Rica

The process of economic development in Latin America in general and Central America in particular has been characterized by a heavy emphasis upon industrialization and agricultural exports. This process has been one of the determining factors in the distribution of wealth and resources; property rights have been historically biased toward powerful minority groups. Therefore, land tenure has been at the center of the political stage for many decades.

Central America is facing an average population growth rate of 3.5 percent. The region has over 20 million inhabitants, with more than 52 percent in the rural areas. FAO estimates a per capita income of US$ 400 in this sector. Table 1 presents the population density by countries. The worst situation is found in El Salvador with a population density of 204/km^2 and a vigorous migration to the urban centers. The other countries in the region, although not as critical, are also starting to feel population pressure.

Table 1. Population density in Central America. Source: FAO 1981.

Country	Area (km)	Population (000s)	Population Density (people/km^2)
El Salvador	20,965	4,255	203
Guatemala	108,888	6,621	61
Costa Rica	50,899	2,111	41
Honduras	112,087	2,831	25
Panama	75,548	1,826	24
Nicaragua	148,005	2,395	16

Within the countries of the region land and capital (or access to credit) are notoriously maldistributed, as are infrastructural services, access to technology, education, etc. The combination of these factors determine the land use pattern for any given production activity. Moreover, at the macro level these factors facilitate the characterization of a set of categories of farmers with specific production constraints and objectives.

The resource distribution problem poses more than just an ethical problem; it has also become a practical one. It is necessary to equate society's long-term welfare objectives with its immediate sociopolitical needs. In particular, the process of economic development, with its inherent maldistribution of resources, has forced a detrimental pattern of land use, whether the responsible parties represent small farms, large farms or multinational corporations.

The plundering of the resources has delineated the most important ecological change taking place in the region: *deforestation*, a direct result of the accelerated rate of conversion of forest trees into grass land or cropland. During the last three decades in Central America, two thirds of the area under forest has been lost to other types of activities. Moreover, the rate of deforestation has increased at a faster pace during the last decade. FAO has estimated that 15 percent of the Central American forest has been lost during this period, i.e. some 26,000 km^2, mostly from Costa Rica and Nicaragua. Table 2 summarizes the loss of forest area in each country.

Table 2. Decrease in forest area by country, 1970-80. Source: FAO.

Country	Area lost $(km)^2$	Percentage of Total Forest
Costa Rica	7,370	29
El Salvador	400	22
Nicaragua	11,400	20
Honduras	8,200	17
Guatamala	5,500	11
Panama	3,000	7
Belize	350	3
Total	36,220	15

Table 3 presents the relationship between the farm size and the percentage of farm land allocated to forestry activities in Honduras. Farms in the smallest stratum (0-5 ha) utilize on average less than 1.5 percent of their land for forestry, while the higher strata allocate 8, 12 and 78 percent to forestry related activities, respectively. Moreover, a characterization of the type of forestry activities developed by the different size strata reveal significant differences among them. These are depicted in Figure 1.

Table 3. Percent of farm land in Honduras in different land use categories, by farm size. Source: National census, 1974.

Farm Size	Annual Crops	Perennial Crops	Pastures	Forest
	Percent of total in land use category			
0-5 ha	38	16	2	1.5
5-20 ha	29	28	11	8
20-5 ha	14	18	17	12
>50 ha	19	38	70	78

The type of agroforestry activity practiced on farms in each stratum is a function of available land and the farmer's basic objectives. As the resource endowment of the farm increases, there is a tendency toward specialization in production.

It appears that agroforestry systems could be an invaluable aid in improving the risk aversive strategies of farmers in the short run and sustaining production levels in the long term. The potential of agroforestry is born out of its own merits, as a result of the evaluation of the needs of farmers themselves. Agroforestry systems have already been modeled to be efficient and viable under the farmers' ecological and socioeconomic circumstances. Furthermore, they usually permit the farmers to produce in a sustained fashion, thus also protecting the resource base.

(1) Larger commercial farms

(2) Smaller subsistence farms

Figure 1. Some agroforestry systems in Central America.

The presence of trees as a component of any agroforestry system ensures several direct and indirect benefits, the major one's being: 1) soil conservation by protection of slopes and improvement of soil structure through incorporation of a larger volume of organic matter, 2) possible nitrogen fixation, 3) more intensive use of the farm land through use of marginal land and/or better utilization of the existing resources, 4) satisfaction of most of the family needs for timber, firewood, fence poles, construction material, fruits, etc., 5) diminished risk through product diversification, and 6) increased income and savings.

Development of dynamic agroforestry alternatives for farmers in the Comayagua Valley of Honduras

These basic ideas were utilized to develop a technological alternative in the Valley of Comayagua in Honduras, a fairly flat area of 38,000 ha at 625 m above sea level. Most of the farms in the project area fall in the 5-20 ha size category. Their productive activities are integrated systems with crop, animal and forestry components. The rainfall varies between 800 and 1200 mm/year and is concentrated in the period from May to October, with a dry season from November to April.

The working hypothesis behind the agroforestry intervention was: An appropriate agroforestry system can indeed facilitate changes in land tenure at the farm level, or at least it can strengthen the existing property rights.

The agroforestry system use is based upon a set of characteristics determined by the ecological zone, as well as by the objectives and the economic and structural constraints of the clientele. Thus, the land tenure situation was the first determinant of the type of productive system developed. In this particular case the project's clientele own small areas of land in a region with a prolonged dry season. Their resource endowment (land, capital and labor) would place them in the intermediate group. That is, their primary objective is subsistence production and they occasionally sell their production surplus. These factors were the basis for defining the characteristics of the production alternative developed. Accordingly, the agroforestry system should:

1) be able to bring sustained increases in productivity;
2) aid in augmenting income stability;
3) minimize risk via an increase in productive diversity;
4) be simple and fairly costless;
5) maximize the number of interactions among components of the production system.

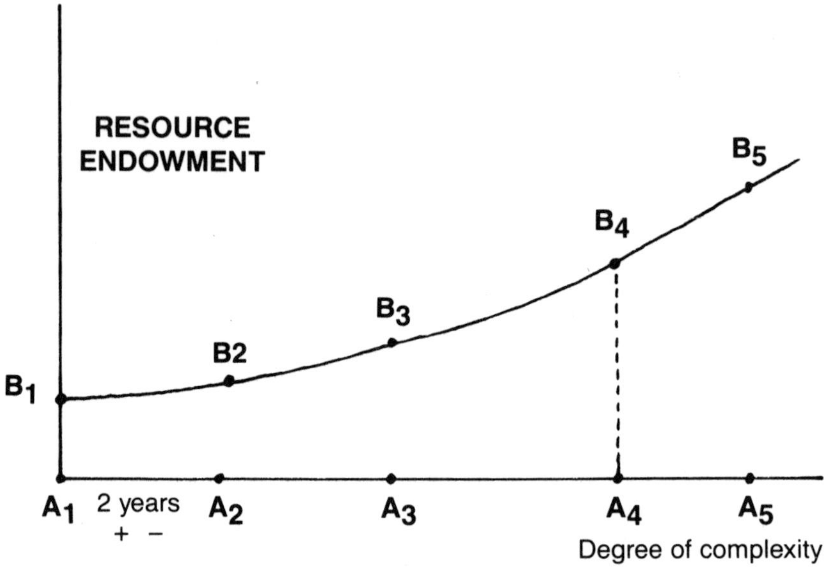

Figure 2. Phases in the development of a dynamic agroforestry alternative in Comayagua Valley, Honduras.

Phases

1	$A_1B_1 =$	Traditional system (collective land use)
2	$A_2B_2 =$	Use of leucaena and sugar cane during dry season, release of pressure on collective land
3	$A_3B_3 =$	Use of leucaena and sugar cane. Use of *Erythrina poeppigaina* as living fence, fuel and small animal feed. Use of some fruit trees such as *P. guajaba*
4	$A_4B_4 =$	The same as phase 3, plus the use of specific area to grow commercial trees such as *Cordia alliodora* or other species adapted to the region
5	$A_5B_5 =$	New phases might follow

The agroforestry system designed on this basis allowed the strengthening of the farmers' property rights. As a result of income improvements productive investment on the farm was enhanced. This was the case of "forestation."

Under the traditional system the small farmers allocate 7 hectares for corn and sorghum and 11 hectares for natural pasture to sustain their cattle herd (see Figure 2). However, they need to move the herd to collective and/or government land during the dry season. Although this land is a fairly large area (approximately 1000 ha.), it becomes overgrazed. Consequently, the herd loses weight and practically stops its milk production.

During phase 2 the farmers are given incentives to introduce 0.5 ha of *Leucaena leucocephala* and sugar cane, which leads directly to a change in the land use pattern. That is, some of the area(1 ha) with maize and sorghum is allocated to the production of leucaena and sugar cane. This change is implemented to satisfy the cattle feed requirements and in some cases leads to a more intensive use of land, since some previously unused marginal land enters into production. This change in the land use pattern has an immediate secondary impact: the release of some of the pressure on the collective land.

The third phase is initiated by planting *Erythrina poeppigiana* as living fences and also some fruit trees to enrich the existing garden plot. The fourth phase involves the "commercial" tree planting stage; that is, the allocation of unused or marginal land to the production of trees for sale. This should be easy to implement if there is no competition between present production activities and the new ones. Furthermore, once the small farmer has sensed the productivity and financial differences brought about by the changes implemented during the earlier phases, it becomes easier to implement this "commercial" stage. The time elapsed between phases is a function of the ecological and socioeconomic circumstances. A time interval of 2 years between phases is used for this example.

References

Budowski, G. 1982. The socio-economic effects of forest management on the lives of people living in the area: the case of Central America and some Caribbean countries. In E.G. Hallsworth (ed). *Socio-Economic Effects and Constraints in Tropical Forest Management.* Wiley. Chichester.

CATIE. 1984. Caracterizacion ambiental y de los principales sistemas de cultivo en fincas pequenas: Comayagua, Honduras. Informe Tecnico No. 42. CATIE. Turrialba, Costa Rica.

Gutierrez, A.W. 1983. Caracterizacion de los sistemas predominantes en fincas familiares de Cariari y Monteverde, Costa Rica. M.Sc. Thesis. CATIE. Turrialba, Costa Rica.

Peck, R.B. 1979. Traditional forestation of local farmers in the tropics. In G. de las Salas (ed). *Proceedings of the Workshop on Agroforestry Systems in Latin America*. CATIE. Turrialba, Costa Rica.

Sepulveda, S. 1985. Integrated production systems research methodology. CATIE. Turrialba, Costa Rica.

Land Tenure and Agroforestry in the Dominican Republic

Isabel A. de Ceara

Instituto Superior de Agricultura La Herradura, Santiago, Republica Dominicana

Introduction

In this paper I propose to outline the land tenure systems in the Dominican Republic, describe the agroforestry and silvopastoral systems, and explain how the latter are related to the former. In addition I will present the experiences that rural development and natural resource management programs have had in establishing agroforestry systems. Finally, the paper will point out the key factors that ought to be taken into account when designing agroforestry programs in the Caribbean and Central American region.

Caribbean and Central American countries are facing an urgent need to improve their production systems in order to guarantee food, fiber and energy for their growing populations. The increasing pressure on these countries to produce a greater number of export products, using in many cases imported technology requiring great capital investments, has displaced the small farmer to more ecologically fragile zones. These zones require a more rational exploitation from an ecological and economic point of view.

To group Central American and Caribbean peasants by land tenure systems would be an extremely complex task. Nevertheless we can point to several common characteristics among these systems:

- The greatest number of private agricultural holdings have an average size of one hectare. The greatest amount of land is in the hands of a few large landowners.

- There is a considerable number of peasants which do not own land. Thus when we talk of land tenure we must consider not just owners, but occupants, usufructuaries and sharecroppers.

- Agroforestry systems are a priority for the smallholders. This is not the case for large landowners who are in a position to dedicate their holdings exclusively to commercial forestry exploitation since they do not have the same urgent need to generate income in the short term as the smallholders.

- The concept of agroforestry is utilized in the majority of small holdings, although not as a systematically understood concept. The producer and his family intuitively recognize the benefits generated by agroforestry systems: short cycle or perennial crops for food, animal production, trees for timber and firewood and medicinal or curative plants.

- It is essential to study the systems utilized by the farmers themselves so as to be able to design and introduce improvements through the use of new or improved species.

This paper will focus on the main issues concerning land tenure and agroforestry in the Dominican Republic to illustrate the situation in Central America and the Caribbean.

Land tenure patterns in the Dominican Republic

The distribution of land

The Dominican Republic has 6.2 million inhabitants living in an area of 48,297 km^2 of which 55 percent, or 2.66 million hectares, is comprised of farms or holdings. These 2.66 million hectares are divided into 385,000 properties. Eighty-five percent of these properties have less than 5 hectares occupying 12 percent of agricultural lands: the minifundia or smallholdings. In addition, 16 percent of the farmers work in holdings with less than 0.5 hectares. Eighteen percent of the producers own 88 percent of the lands forming in this way the well known bimodal land distribution pattern of "minifundia-latifunda" so typical of the Central American and Caribbean region.

Between 1971 and 1981 the number of holdings with less than 5 hectares has increased, which indicates a tendency towards greater fragmentation of property. This phenomenon has not been as marked in holdings of 5 to 1000 hectares. See Table 1.

In 1971, 71 percent of the holdings were legally farmed while one percent of the land was rented and 10 percent illegally farmed (Republica Dominicana, 1971). The largest number of non-owned small farms fall into the categories of "occupants or free beneficiaries" and "occupants under concession." Most of these are squatters on government land or on large private holdings; they do not have to pay any kind of rent or tribute. Sharecropping is the next most prevalent form of tenancy, followed by rented land (Barraclough 1970).

The control of the irrigated lands is in the hands of large landowners. Nine percent of the large holdings had, in 1971, 64 percent of the irrigated land while 91 percent of the smaller farmers controlled only 36 percent of the irrigated lands (Barraclough 1970).

The strategies directed to establish agroforestry systems must be designed according to the size of the holding worked by each farmer.

Short cycle crops are planted mostly in small farms. Those with more land dedicate it to sugar cane, forestry or grazing. The smallholdings are producers of rice, cacao, coffee, tobacco and roots and tubers. The land is used more intensively in the smaller holdings. The amount of land per worker ranges from 0.27 hectares in the small units to 1.54 hectares in the family size units to 13.2 hectares in the multi-family large holdings.

The distribution of the land influences the type of crop planted more than the form of tenancy. Though there are no detailed studies on cattle lands, it is known that the great majority of large landowners dedicate their holdings to raising cattle, and in some instances goats, without applying any type of controlled management. Sixty-one percent of the cattle properties have herds comprised of less than 6 head of cattle and 14.7 percent of these farmers (of which 85 percent have holdings of less than 5 hectares) have only one cow. The majority of herds consisting of 2 to 5 head of cattle also belong to producers with less than 5 hectares. The large land owners specialize in raising beef cattle. For example, 69 percent of the farmers owning 200 to 500 hectares possess beef herds and only 18 percent of those farmers with less than 5 hectares raise cows (Republica Dominicana, 1982).

This concentration of land and resources in the hands of a few producers and the growing need to answer the claims for increased social justice put forth by the peasants themselves has led to land distribution and redistribution schemes, or agrarian reform, in almost all Central American and Caribbean countries. The reforms of the land tenure systems have as their prime objective to provide for the peasant and his family, access to the main means of production, land. The forms and strategies used in land reform differ according to the political systems and social movements prevalent in each country.

Table 1. Number of holdings, surface, and average size according to the size of the holdings. Source: National Agricultural Census 1971 and 1981, Oficina Nacional de Estadistica.

	National Agricultural Census, 1971		/	National Agricultural Census, 1981		
Size of holdings (hectares)	Number of Holdings	Total Surface Area (hectares)	Average Size of Holdings (hectares)	Number of Holdings (hectares)	Total Surface Area (hectares)	Average Size of Holdings
Total	304,820	2,719,305.6	9.0	385,060	2,659,977.4	6.9
Less than 0.5 hectares	49,651	12,132.0	0.25	61,670	12,465.0	0.18
0.5 - 5 hectares	185,292	337,516.8	1.8	252,995	311,660.0	1.25
5 - 10 hectares	33,803	229,930.1	6.8	32,543	230,383.0	7.0
10 - 50 hectares	28,987	584,165.0	20.1	30,815	636,484.2	20.6
50 - 100 hectares	3,974	266,841.0	67.1	4,081	270,154.0	66.1
100 - 200 hectares	1,791	247,245.2	138.1	1,825	249,629.0	136.8
200 - 500 hectares	884	266,350.1	301.3	786	230,175.3	292.8
500 - 1,000 hectares	222	146,861.0	661.5	184	120,574.9	655.3
1,000 or more hectares	216	628,264.4	2,908.6	161	598,452.0	3,717.1

Agrarian reform in the Dominican Republic

In the majority of Latin American countries the modern history of agrarian reform movements had its origins in the international pressures brought about by the Alliance for Progress programs pushed by the United States government beginning in 1961. Nevertheless, in the Dominican Republic the elimination of 30 years of tyranny had more to do with the rise of the agrarian reform process than international demands. (Instituto Superior de Agricultura, 1983)

The Dominican Agrarian Reform has sought to achieve four basic objectives:

1) To protect the frontier from the growing encroachment of Haitian nationals in border lands. This program was one of colonization and beneficiaries of these colonization schemes along the frontier were called "colones" or colonizers. The economic and social impact of these programs was minimal.

2) To act as a mechanism for creating a political support base. In order to institutionalize this process after the fall of the dictator the Dominican Agrarian Institute (IAD) was created in 1962.

3) To act as a mechanism to arrest the increasing social unrest in the countryside due to the large number of landless peasants.

4) To act as a mechanism to assure self-sufficiency in rice without increasing the price of this grain.

Until 1972, all agrarian reform settlements were comprised of individual holdings. The state distributed properties between 3.13 and 4.69 hectares on average. The beneficiary was only granted a provisional title and had the right to the usufruct of this property but could not sell or transfer the property rights.

After 1972 a new form of agrarian reform settlement emerged, the collective. These settlements were under the direct control of IAD and the beneficiaries working in them were to divide the net income in equal parts. In the collective, the beneficiary's family can inherit the right to the usufruct in case of his abandonment of the plot or his death. In the beginning, this type of settlement applied only to rice lands. After 1979 all land distributed by the IAD was in the form of collectives.

At the moment, a third model for agrarian reform settlements is emerging, the associative. This model permits the beneficiary the right to manage his own parcel in association with other reform beneficiaries in the settlement. The farmers form an association which has the legal rights to obtain credit, contract machinery, and obtain other inputs from independent providers. In general, each beneficiary works only his own land with the help of his family. The benefits derived correspond to the production of his parcel. This reform model shows the great organizational capacity that exists among the peasants.

The associative model has emerged in two forms, depending on whether the settlement in question was formerly an individual or collective agrarian reform settlement. Congress has yet to approve the law which would give IAD the right to legally adjudicate land to associations.

Table 2 shows the total number of agrarian reform settlements established, the number of families benefitted and the total amount of land distributed by IAD.

The forestry situation in the Dominican Republic

Ten to fifteen percent of the forest land in this country is privatedly owned and extensive areas of deforested land exist under no legal ownership status. These lands are used for migrant agriculture and extensive dairy or beef production. The increasing rate of the deforestation process in the country has been of such magnitude that, although in 1947 it was estimated that 60 percent of the national territory was covered by forest at the present time only 26 percent is covered by forest, including species in dry forests (areas with an annual rainfall below 1,000 mm). Much of the forest has been destroyed in the process of producing timber, firewood and charcoal. One consequence of this is that almost 300 native species are threatened with extinction. The main river basins are in poor state due to the constant destruction of forest vegetation, the number of dry rivers has increased and there are grave problems with soil erosion.

Table 2. Agrarian reform: settlements, families benefited and land distributed in the Dominican Republic, 1982.

Type of Settlement	Number	Percentage	Families Benefited	Percentage	Land Distributed (hectares)	Percentage
Colonies before 1962	40	9	11,451	19	139,899.8	38
Individual settlements, 1962-1982	267	60	32,284	54	170,894.5	46
Collective settlements, 1972-1982	136	31	15,676	27	61,499.3	16
Total	443	100	59,411	100	372,293.6	100

Source: Pablo Rodriguez, Leo Colon, Juan Ogando, Randy Stringer, y William C. Thiesenhusen. "Agrarian Reform in the Dominican Republic: The Case of Ysura", Case prepared by Proyecto de Colaboracion Institucional entre Instituto Superior de Agricultura-Centro de Administracion del Desarrollo Rural (ISA-CADER), Instituto Agrario Dominicano (IAD), y Land Tenure Center (LTC). Santo Domingo, August, 1983.

The production of woodfuel is about 2,000,000 m^3/year, of which 65 percent is consumed as firewood and 35 percent as charcoal (Tejada 1984). Ninety percent of the firewood consumption takes place in rural areas which generally exist outside the market economy. Of the total timber that is processed, some 80 percent goes to urban homes. The woodlands that exist today have an average yearly production of 10m^3 /ha. This implies that 200,000 hectares are exploited annually to satisfy the demand. In the last 14 years, however, only 3,157 hectares of forest have been established.

The low per capita income in the Dominican Republic (RD$450 to RD$600) lends itself to migrant agriculture because people tend to provide themselves with charcoal, firewood, and other raw materials from the forest. Charcoal making had been a sporadic economic activity, but with an increase in rural demand, it is becoming a full time activity for some farmers. Of all migrant agricultural practices, charcoal making generates the principal income, and agricultural products are mainly for household consumption.

The lack of professional advice and technology, and the land ownership patterns and low income per capita are factors that lead the small farmers to practice migrant agriculture. Migrant agriculture is being practiced here in one of its variant forms: recurrent agriculture, rotating the land under cultivation but living permanently in the same area. The cycle includes tree cutting, charcoal and fuelwood production, cropping, grass planting, abandonment and forest regrowth. The crops grown are beans, potatoes, and pigeon peas. Low land broad-leafed forest trees have been deforested by slash and burn agriculturalists and charcoal makers.

The national forest service, FORESTA, has the responsibility to enforce the 1967 law which closed saw-mills and prohibited the cutting of trees (Olivo Veras, 1984). This forestry law and the 1983 law on utilization of wood produced locally constitute an obstacle to the rational use and establishment of forest species.

The forestry policy of the country should be oriented towards providing:

1. Security, so that what is planted may be harvested;
2. Credit incentives;
3. The establishment of agrarian reform agroforestry settlements in those lands suitable for forestry;
4. Rational exploitation of the natural forests and plantations;

5. The establishment of farms that produce firewood and charcoal, given the fact that a rural family needs 0.33 hectares per year to produce its firewood. It is estimated that a peasant family can manage adequately 10 hectares of *Leucaena leucocephala* and could generate some US$3,000 per year from firewood production (Tejada 1984).

Agroforestry practices and their relationship to land tenure systems

Agroforestry includes the management of agricultural or forage species in combination with forest species in the same plot of land in order to achieve a better use of the soil, water, light, and the farmer's labor in a sustained fashion.

Forestry activity in this country faces a series of obstacles and contradictions: the farmers immediate needs, the spatial distribution of benefits, the lack of an agricultural culture that fits the environment and incentive policies (Santos 1984).

The farmer's immediate needs

Forestry activities differ from agricultural activities in that the return on investments and the flow of benefits occur in the long term. But in general, people tend to place little value on income to be gained in the distant future.

The income yielding capacity of a forest plantation will oscillate between 6 and 20 percent annually. But capitalists wish to invest in economic activities whose return on investment occurs within a fairly brief period of time. In order for capitalists to find it attractive to establish a forestry enterprise they will need to receive incentives which increase the rate of return on investment and assure a secure and stable future.

For a peasant toiling in the mountains, the argument that a hectare of pine trees will produce in 20 years one hundred times more than that same hectare will produce today if he plants beans is meaningless. His basic needs are immediate and a stream of income occurring 20 years from now cannot satisfy those needs today. Thus for him the present value of this stream of income is zero. Given his present conditions the peasant cannot afford to wait. The peasant's strategy is to plant now and plant crops whose cycles will be as short as possible.

The pragmatism of the capitalist and the farmer argues against an ecologically optimal use of nature.

In order to modify this situation one must offer financing which makes it affordable to wait (for the capitalist) and redistribute land in such a way that each family has enough land to sustain itself above a level of absolute poverty. It will require the design and transfer of a technology appropriate to the environment, so that the peasant family may produce its own food and fuel without harm to the soil. The hope that is placed on agroforestry systems to solve this problem is great, but in order for these to work one must first determine the income yielding capacity needed to motivate the capitalist to invest, and develop compatibility between short cycle crops and a healthy ecological environment to motivate the peasant.

The spatial distribution of benefits

The spatial distribution of benefits also affects the establishment of permanent crops. The people who, in order to protect the dams and rivers, must change their behavior *vis-a-vis* nature are not the same people who will benefit from such a change.

The increase of productivity achieved through irrigation, the greater availability of electric energy produced at lesser costs and the improved operation of aqueducts are without exception benefits received by those who live downstream, in the valleys. On the other hand, the change from short-cycle to permanent crops (the sacrifice in waiting this entails) and the soil conservation works that consume so much human energy must be undertake by those living on the hillsides. It is possible that agroforestry systems may redress this imbalance by allowing those on the hillsides to combine their short cycle crops with permanent crops and by forcing those living downstream to compensate the hillside inhabitants through low cost credit, distribution of land and plants, health, education, roads and other public works.

Lack of agricultural culture

Another obstacle facing forestry activities is the lack of an agricultural culture that fits the environment. Historical factors in many countries have forced the immigration of inhabitants of flatlands to the hillsides. Since one cannot hope to relocate all hillside inhabitants in the flatlands we will have to create a generation of forest agriculturalists through schooling and technology transfer.

Policies and political aspects

The political aspects of our country also have great influence. Given the fact that democracies are far from institutionalized in the Third World, to sustain them is a constant struggle. Governments fight for their own survival and thus their activities (including investments and expenditures) are directed towards maintaining themselves and the system in power. Long term investments whose benefits will not be forthcoming during the term of office are unlikely to receive top priority. In addition, credit and technical assistance policies tend to favor the establishment of single crop systems of short cycle.

Agroforestry is faced not just by technical or logistical problems. Reforesting is above all a problem of lack of adequate survival and exploitation techniques for hillside inhabitants. It is a problem of subjectively low income yielding possibilities given time lapse and uncertainly factors. Finally, it is also a matter of accepting that it is impossible to stop the deforestation process without solving at the same time the problems of the inhabitants of these areas.

Various forms of agroforestry exist in the Dominican Republic. However, these have neither been systematically established nor have their efficiency and profitability been thoroughly verified. The agroforestry and silvopastoral schemes most commonly found in this country are listed in Table 3.

A study conducted by Mercedes (1980) indicates that hills with slopes of up to 75 percent could support fruit crops such as *Anacardium occidentalis* (cashew), *Annona spp.* and *Mangifera indica* (mango). According to the author's scheme, shrubs and herbaceous species would be part of the second canopy and used for grazing. Hills whose slopes fall between 24 and 74 percent can be used for producing forest crops of poles, firewood and charcoal.

Herbaceous species and shrubs are used to protect the soil against erosion and provide forage. Agricultural crops can be grown between the rows (corn, beans, pigeon peas, pineapple, tomatoes and eggplants).

Table 3. Some of the most important agroforestry systems in the Dominican Republic. After Rafael Fajardo (1985).

Location	System	Tree Component	Agricultural Component	Products of the System
All throughout the country (35,000 km live fencing)	Agrosilvopastoral	*Gliricidia sepium* *Guazumaulmifolia* *Prosopis juliflora*	Pasture various crops	Posts, firewood, foodstuff (for human and cattle)
San Jose de Ocoa San Juan de la Maguana	Agroforestry	*Inga spp.* *Erythrina spp.*	*Coffea arabica* Musaceas	Shade, soil protection, organic matter, industrial grain, food-stuff
La Vega, San Fco. de Macoris	Agroforestry	*Erythrina spp.*	*Theohoma cacao* Musaceas	Industrial grain, food-stuff, shade and soil protection
Bonao	Agroforestry	*Cocos nucifera*	*Coffea arabica* *Theohoma cacao*	Industrial grain and nuts for human consumption, soil protection
La Vega, Nizao	Agroforestry	*Mangifera indica* *Sabal umbraculifera*	*Zea mays*	Wood, fiber for artesany, leaves for roofing, soil protection, fruits, grain for human and animal consumption
Pariso, Polo	Agroforestry	*Citrus spp.*	*Coffea arabica* Musaceas	Fruits for industry and food, shade, soil protection
Puerto Plata	Agroforestry	*Caralpa longissima* *Persea americana* *Mangifera indica*	*Theohoma cacao*	Shade, soil protection, fruits for human consumption and industrial grain

Bani, San Juan de la Maguana	Silvopastoral	*Prosopis juliflora* *Sabal umbraculifera*	Pasture	Shade, forage, food for cattle, fiber and wood for housing, soil protection
Bani, San Jose de Ocoa	Silvopastoral	*Guazuma ulmifolia*	Pasture	Shade and cattle feed, soil protection
Bonao, Puerto Plata, Sosua, Gasper Hernandez, Rio San Juan	Silvopastoral	*Roystonea hispaniolana*	*Digitaria decumbens*	Shade, fiber and wood for rural housing, fruits and pasture for cattle, and protection
Sosua, Gaspar Hernandez, la Romana, Hato Mayor	Silvopastoral	*Citrus sinensis*	*Digitaria decumbens*	Shade, fruits for human consumption, pasture for cattle
Sosua	Silvopastoral	*Coccolobis uvifera*	*Digitaria decumbens*	Shade, cattle feed, soil protection
Sosua	Silvopatoral	*Swietenia mahogani*	*Digitaria decumbens*	Shade, cattle feed, soil protection and wood for construction
San Pedro de Macoris	Silvopastoral	*Bursera simaruba* *Guazuma ulmifolia*	Natural pasture	Posts for fence, soil protection, shade, cattle feed
Zonas Montanosas	Silvopastoral	*Pinus occidentalis* *Pinus caribaea*	*Melinus minutiflora*	Soil protection, wood, pasture for cattle, resin, paper pulp
Bosque Seco	Silvopastoral	*Prosopis juliflora*	*Panicum maximum*	Soil protection, firewood, shade forage and food for cattle

Farmers who own land or who have occupied state lands without title utilize an agroforestry system consisting of fruit trees, coffee, cacao and citrus. Some agrarian reform settlements also use this system. Nevertheless, the temporary occupant of private lands will hardly dedicate his time to this type of exploitation, which requires permanency and security. Farmers and peasants with small holdings cannot use the system at all, since it conflicts with the primary need to assure their survival in the short run.

The system of coniferous trees or *Gliricidia* on pasture lands is commonly used by medium to large holders who do not face the same urgent need to plant short cycle crops. The grazing lands in the lowlands are surrounded by *Gliricidia sepium* as a live fence.

There also exists a type of agroforestry system specific to sharecroppers. In the majority of cases sharecroppers must share their harvest with the owner. Yet in some parts of the country the sharecropper only has the responsibility to clear the forest, cultivate it for his own good and at the end of the contract return the land to the owner planted with pasture. In these cases it is very difficult for the sharecropper to establish an agroforestry system since in the long run he will not be able to benefit from this effort and his contract demands that he return a one crop plot (pasture).

Farmers who occupy private land are generally less permanent than those who occupy state-owned lands, and they will be less disposed to establish permanent crops. Yet in the case of sharecropping, leasing, or illegal occupation, the owner must pay for the improvements the farmer has brought to the plot, including trees planted. Some farmers could then use these permanent crops as a way of claiming rights to payment for improvements.

In the new forestry plantations belonging to the state, agroforestry systems are not at all considered. The interest of the government in these lands is to keep the farmer as far away as possible in order to preserve the forest.

The management of grazing goats and beef cattle is being studied in plantations which produce firewood and charcoal so as to determine what species used for firewood and forage permit ample grazing without undergoing total destruction or what species have a high degree of regeneration or resprouting capacity. The most immediate concern of these studies is to determine the optimum number of grazing animals and the optimum grazing time for the pasture.

Existing forestry projects and their agroforestry components in the Dominican Republic

Plan Sierra

Plan Sierra is a rural development program which began operating on April 1, 1979. The objective of this program was to improve the degraded condition of the forest and soil in an area where 14 rivers, the source of water for the northeast region of the country, are born. It also sought to relieve the state of critical poverty in which the majority of the inhabitants of the region lived. It hoped to introduce techniques that would allow the inhabitants of the Sierra to live with nature without destroying it. Supported by the Ministry of Agriculture, Plan Sierra has received donations from the W.K. Kellogg Foundation, the Swedish Government and other international institutions.

The owners or occupants of the Sierra can be divided into two groups: poor peasants who barely produce enough to feed their families, and peasants or large landowners who can afford to invest in production. The former group has no possibility to reforest since they cannot afford to dedicate the land they use to produce food for their families to planting trees which take 15 or 20 years to yield benefits, even if these trees are planted by volunteers free of charge.

On the other hand, members of the latter group consider that it is far more profitable to dedicate their lands to extensive cattle raising, or even let them lie fallow, than to plant them with trees for which they would need a minimum investment of RD$800/ha in the first three years, which could not be recovered for the next 15 to 20 years.

The only means of stabilizing the ecosystem is to stabilize the social system. To solve the reforestation problem one must solve the poverty problem. The distribution of land in Plan Sierra has led to the design of two models of reforestation. These are presented in Table 4. Reforestation models which affect neither the distribution of land nor the land tenure system have two fundamental weaknessess: one is that the socioeconomic solution to the plight of one particular family depends in large measure on the amount of land at that family's disposal; the second weakness is that reforestation depends on the will of the land owners and not in the objective necessity to reforest that particular tract of land (Republica Dominicana 1983).

Table 4. Models of forestry production according to the system and distribution of land. Plan Sierra, Dominican Republic, 1984.

A. Models which affect the tenancy and the distribution of land

A.1. For each family of 6 members with 3 hectares of land (50,000 hectares)

a) 0.062 ha for orchard
b) 0.625 ha for food for self-consumption and grain to feed 10-15 chickens
c) 1.75 ha for permanent crop (i.e. coffee)
d) 0.25 ha for forage to sustain one cow
e) 0.31 ha for forest to obtain firewood

A.2. Organization of Forestry Enterprises (112,500 hectares) These enterprises will be state owned and would utilize peasant labor. Model 1 and 2 are complementary.

B. Models which affect neither the distribution nor land tenancy

B.1. Each family according to the extension of its land will be given assistance to establish a model A. 1. farm. If the family has more than 3 ha the objective will be motivate it to plant wood producing species.

B.2. Contract families to reforest 19-25 has in return for a salary and for helping it establish model B. 1.

B.3. Reforestation of lands given by the owners with paid labor with the legal compromise of reimbursing Plan Sierra the costs once the trees have been harvested and sold.

B.4. Nurseries in schools in order to establish school run forests and create a forestry culture.

MARENA

The Management of Natural Resources Project (MARENA), being carried out by the Subsecretariat of Natural Resources, has at its disposal RD$ 21 million. It was started in 1982 and its objective is to increase the income and standard of living of the poor peasant who works on the hillsides, through soil and water conservation and institutional strengthening. It receives technical and financial support from the U.S. Agency for International Development (USAID).

One of the most important areas of influence of the MARENA project is the Ocoa River basin which covers some 500 km^2 and includes about 5,000 producers, most of whom are untitled and have one hectare farms. The project's activities in this area concentrate on conservation, basin protection, investigation of agricultural systems appropriate for hillside farming, and credit incentives for soil conservation and reforestation and improvement of soil and water quality).

The investigation of hillside agricultural systems reveals the use of live barriers using different species: limoncillo, supermerker, alfafa, sorghum and guatemala *(Tripsicum lasum)* all of which may be used as medicinal plants and forage and which also conserve the soil. In addition, strips consisting of one or two rows of *Leucaena leucocephala* are being established for erosion control, forage and fuelwood.

The project has programmed an inventory of the agroforestry systems in the area, but this has yet to be started. At this time the project is busy studying the profitability of plots where fruit trees (figs) are planted in combination with short cycle crops (beans).

One common type of agroforestry system in the area is the cultivation of potatoes *(Solanum tuberosum)* with *Pinus occidentalis*. This system has not been studied. In the majority of cases it is practiced in lands of very pronounced slopes (over 35 percent), suitable for forestry. Though this system is widely practiced it should not be tolerated or encouraged since the region's topography requires permanent crops and not crops which, like the potato, cause the soil to be overturned during harvest time.

MARENA is also considering establishing species in farmers' lands which can be used to meet energy needs.

Wood as Fuel Program

The Wood as Fuel Program is a component of the Project for the Utilization of Energy Resources, presently being directed by the National Commission for Energy Policy with USAID financing. The program has been implemented by the Instituto Superior de Agricultura (ISA) beginning in 1983 in a 1,000 hectare dry wood forest in Mao.

The Wood for Fuel Program's main objective is to study the native species of the dry forest (where the annual rainfall is less than 1,000 mm) which have good energy use potential. The program also investigates plantation management systems and searches for the best methods to convert firewood into charcoal.

One of the Program's interests is to initiate agroforestry investigations. In particular, it hopes to determine which dry forest species are suitable for use as goat feed and as firewood. Goat raising in private or large state owned land is a common practice of small producers. Overgrazing consumes the species suitable for use as firewood and accelerates the impoverishment of the soil. The program wishes to determine the best species for grazing and firewood, the optimum number of grazing animals per hectare and the optimum rotation period.

Table 5 shows the dry forest species with the best potential for use as firewood and charcoal. In addition, the resprouting capacity of each species is indicated. This capacity for resprouting, a characteristic of certain species, permits the continuous use of the plant without requiring great investment or high labor costs. *Gliricidia sepium* is one species which, in addition to being an excellent source of firewood and charcoal, also has great resprouting capacity. However, this species is not indigenous to dry forests.

The planned or established experiments in agroforestry include:

1) Plantations of *Azadirachta indica* and *Leucaena leucocephala* with interplantings of beans and corn. Another experimental plot in Santiago, with *Eucaliptus citriodora* and beans in the first year, saved considerable money.

2) Silvopastoral experiments on ISA pasture land:
   a) Forage tree species for grazing and cutting;
   b) Controlled grazing beneath plantations of different spacings.

Table 5. Dry forest species in the Dominican Republic with potential use as firewood and charcoal. Source: Programa Madera como Combustible, Instituto Superior de Agricultura.

Common Name	Scientific Name	Resprouting Capacity
Guayacan	*Gauicacum officinale*	+
Cambron	*Prosopis juliflora*	+
Guatapanal	*Caesalinia coriaria*	+
Sangretoro	*Maytenus buxitolia*	
Candelon	*Acacia scleroxyla*	
Quina	*Exostema caribeaun*	+
Brucon	*Cassia emarginata*	
Mastozo	*Capparis flexuosa*	
Baitoa	*Phyllostylon brasilienses*	+
Tabacuela	*Pictetia spinifolia*	
Uvero	*Coccoloba leoganensis*	
Campeche	*Haematoxylon campechianum*	
Cogiamo	*Krugiodendron ferrum*	
Almacigo*	*Bursera simaruba*	+

*The Álmacigo has more important uses besides firewood. Its quality as firewood is not very good.

3) Grazing trials as a means of controlling weeds and competition in native dry forests.

4) Enclosures to determine impacts of goats on heavily grazed dry forest at ISA.

Growth rates of the dry forest are not known, due to the heavy grazing and harvesting pressure in virtually all of its extent. One of the values of the large forest research property that ISA maintains in Mao is that the area contains about 500 hectares that are to be preserved in natural condition, without livestock and with some experimental silvicultural plots.

The state and privately owned lands consisting of dry forest are utilized by communities to extract firewood and graze goats and cattle. As the management of old and new plantations is improved, community access to these forests will be limited, forcing farmers to buy firewood or to produce it in their own plots.

The scarcity of firewood for domestic use or for making casabe (a bread made out of *Manihot esculentum*) represents an additional burden which must be born by thousands of peasant families. Rural women, more than any other members of the family, will suffer from the lack of fuel. At this time a peasant roams through about 2.2 kms daily in order to obtain firewood to prepare his meals (Bueno and Checo 1985). The distance has increased, which indicates that sources of firewood near the home are being exhausted (Hoskins 1980).

Rural communities which do not have firewood at their disposal must buy it. Charcoal represents a major cost for these families, since gas is used as fuel mainly by urban families. I estimate that 13 percent of a farmer's basic income is used in buying energy to cook his meals. Some women's associations have asked for the establishment of communal forests as a source for firewood for domestic use and for making casabe. Agroforestry systems represent a unique way of providing firewood and food, especially if forage species can serve as firewood.

Communal forests represent an option. Nevertheless, the state, though it may donate land for this purpose, would have to regulate access. This arrangement could be established in collective and/or associative land reform settlements for the use of the beneficiaries.

Final observations

Although land tenure systems in Central America and the Caribbean are characterized by private property, a great number of peasants are still landless. Landless farmers go on to be occupants of private holdings or state lands, sharecroppers or leasing lands when their income allows it. Landless peasants represent a constant pressure on social and agrarian reforms. The tenancy structure established by these reforms grants usufruct rights to the farmer, but not the right to sell his plot.

Even if an efficient agrarian reform system existed, there would not be enough land to give all farmers a sufficiently large plot to allow him to satisfy his basic needs and generate a surplus. Thus the survival of those who live in the countryside and of the natural resources themselves so that the continuity of production is assured, will require that special attention be paid to the system of property rights, the distribution of this property and the geographical location of the same.

Agroforestry systems in conditions of extreme fragmentation of property are exceedingly difficult to establish and sustain. In these cases the establishment of communal forests is required.

The experience of rural development programs in the Dominican Republic show that the distribution of the land plays a more important role than property rights in establishing agroforestry systems. Small farmers or peasants who only have a small plot with which to provide for their families cannot dedicate their lands to long cycle crops, except in fence rows along the borders. The medium-sized producers are more susceptible to the introduction of conservation practices. Finally, large landowners, generally absentee landlords, have neither the financing nor the time to care for short cycle plantations and prefer to apply extensive farming practices which generally require low capital investment.

References

Barraclough, S. 1970. Unpublished memorandum. Dominican Republic.

Bueno, S. and H. Checo. 1985. Analisis del consumo energetico de familias pobres en una comunidad rural. Unpublished thesis. Universidad Catolica Madre y Maestra. Santiago.

Fajardo, R. 1985. Descripcion de algunos sistemas agrosilvopastoriles en la Republica Dominicana. *Recursos Naturales* 1:32-38.

Hoskins, M. 1980. La silvicultura comunitaria y las mujeres. *Unasylva* 32(130):27-32.

Instituto Superior de Agricultura. 1983. *Nota Tecnica la Reforma Agraria Dominicana*. Centro de Administracion del Desarrollo Rural. Santiago.

Mercedes, J. 1980. Estudio para el diseno de un sistema de manejo integrado en areas de bosque seco en republica dominicana. Unpublished thesis. Universidad Catolica Madre y Maestra. Santiago.

Olivo Veras, J.G. 1984. Indice cronologico de la legislacion sobre asuntos forestales en la Republica Dominicana. In Universidad Catolica Madre y Maestra, Plan Sierra and Instituto Superior de Agricultura. *Foresta: Alternativa de Desarrollo.* Universidad Catolica Madre y Maestra. Santiago.

Republica Dominicana. 1971. *VI Censo Nacional Agropecuario 1971.* Oficina Nacional de Estadistica. Santo Domingo.

Republica Dominicana. 1982. *VII Censo Nacional Agropecurio 1981: Cifras preliminares.* Oficina Nacional de Estadistica. Santo Domingo.

Republica Dominicana. 1983. Resumen de los primeros 45 meses de ejecucion del Plan Sierra: Informe provisional. Secretaria de Estado de Agricultura. Santo Domingo.

Santos, B. 1984. Foresta: Un problema social. In Universidad Catolica Madre y Maestra, Plan Sierra and Instituto Superior de Agricultura. *Foresta: Alternative de Desarrollo.* Universidad Catolica Madre y Maestra. Santiago.

Tejada, R. 1984. El potencial economico de las fincas energeticas. In Universidad Catolica Madre y Maestra, Plan Sierra and Instituto Superior de Agricultura. *Foresta: Alternativa de Desarrollo.* Universidad Catolica Madre y Maestra. Santiago.

Land Tenure and Agroforestry in Haiti: A Case Study in Anthropological Project Design

Gerald F. Murray

Department of Anthropology, University of Florida

Background

Having done anthropological fieldwork on the evolution of Haitian peasant land tenure, I was invited by USAID/Haiti in the late 1970s to carry out research exploring possible linkages between land tenure variables and the failure of most tree planting projects to motivate Haitian peasants to plant trees. I was specifically asked to assess the degree to which land tenure insecurity served as the principal disincentive to peasant tree planting.

Problem diagnosis

My investigation of some 15 reforestation and soil conservation projects produced the following suggestions:

1) In most project regions the land tenure situation was characterized by the same internal heterogeneity that is characteristic of most of rural Haiti: mixes of inherited plots under individual control, purchased plots, agriculturally marginal land owned collectively by kin groups, State owned land, sharecropped plots, rented plots, and others.

2) Though tenure on some categories of land was insecure, tenure on owner operated plots was quite secure. Peasants regularly invested several hundred dollars (the equivalent of household annual income) in the purchase of such plots. Projects failed for other reasons; and "land tenure insecurity" should not be used as a whipping boy to deflect blame for project failure away from its proper objects.

3) Reforestation projects had traditionally emphasized State land or agriculturally marginal land as the proper site for tree planting. This policy appeared misguided both ecologically and anthropologically. Ecologically the land least "at risk" to erosion from cultivators was agriculturally marginal land, i.e. these plots were low priority plots on which to carry out the "soil conservation" that was so central

a theme of many projects. Anthropologically a peasant planting trees on such State land or kin owned land has absolutely no right to prevent others from cutting down the trees which he plants. Projects were, in short, targeting the wrong tenure categories of plots.

4) Of equal damage, however, were the *technical* choices which conventional projects were making. Fruit trees (with little market value and marginal nutritional value) were being emphasized over wood trees. Those projects planting wood trees emphasized slow growing species. And above all, projects universally resorted to cumbersome polyurethane containers for nursery production, creating formidable transportation barriers.

5) But the most serious program errors were in the *institutional* domain. Funds were routinely entrusted to Haitian public sector ministries which had proven their inability to implement agreed on projects, or to refrain from reallocating project funds to other agendas. Even technically and economically sound projects stood little chance of success if entrusted to the wrong institutional custodians.

Conceptual reorientation

I proposed several basic conceptual themes which would, I predicted, lead to the voluntary planting of millions of trees.

1) *Emphasize economy, not ecology*. Messages concerning the ecological advantages of trees tend to leave Haitian peasants unmoved. Not only do they already know the ecological advantages of trees, they are less interested in nutrient flows or soil flows than in desperately needed *cash flows* into their domestic coffers. Tree projects must be structured in such a way that the planting of trees shows promise of producing a flow of cash into peasant households. Ecological improvements are best pursued in Haiti, I argued, as *a secondary side effect of behaviors in which the peasant engages primarily for economic reasons*. The most promising theme, in this light, would be *wood as a cash-crop*.

2) *Strive for a linkage between two pre-existing anthropological patterns*. The planting of wood trees could be conceived as an evolutionary linkage between two pre-existing economic habits among Haitian peasants, rather than as a new behavior. They are already cash-croppers--much of their food produce is consigned to the internal market system. They are already wood-sellers--they know that there is a

vigorous market for both lumber and charcoal, and many peasant communities derive income from this market. What is missing is a linkage: why not create a project in which, for the first time in their history, they would cash-crop wood?

3) *Use pre-existing land tenure and crop-tenure patterns.* I argued against trying to introduce new tenure or crop ownership arrangements. If the transition to wood cropping were to be smooth, it would have to take place on the same types of plots (as defined by tenure) as the ones on which peasants grow their current cash-crops. And their ownership, relation to the trees should be the same as their ownership control and harvest rights over the other crops which they plant. Most cropping in Haiti is done on individually owned plots whose crops are owned by the individuals who plant them. This means that projects should avoid, in the beginning, exhortations to plant trees on commonly owned kin-land (which is used currently only for grazing) or on State land. No peasant in his right mind plants food crops on either of these tenure types. And projects should avoid, at least in Haiti, the "village woodlot" model. The peasants would have no more confidence in personal income from a village woodlot than they would from a hypothetical "village food-lot".

Program measures

The above concepts were embodied in a project which adopted the following operational measures.

Technical measures

- The higher commercial value of wood led to a choice of the wood tree rather than the fruit tree.
- Fast growing, rather than slow-growing, trees were used: *Leucaena leucocephala*, *Cassia siamea*, *Azadirachta indica*, *Casuarina equisetifolia*, and *Eucalyptus camaldulensis*.
- Border-planting and intercropping techniques were taught to make wood growing compatible with continued food growing on small holdings.
- Most importantly, seedlings were produced in small containers rather than bags. Our pickup trucks could carry, not the standard 250 seedlings, but some 20,000 seedlings ready for transplanting. And an individual peasant could carry 500 trees.

Microeconomic measures

- A contract was made with peasants: they would agree to plant a specified number of trees on their own holdings. In return they would be given the seedlings for free, they would be full owners of the trees, and would be able to harvest and sell the wood whenever they wanted, without project permission.

- The minimum number of trees was 500. Planted at 2m x 2m (to permit initial intercropping of food), this would occupy only a fifth of a hectare. (The average holding is about 1.25 ha.)

Institutional measures

- Absolutely no financial, logistical, or operational link was established between the project and the Haitian government. The funds were channeled entirely through non-governmental organizations. Major grantee NGOs funneled seedlings, technical assistance, and data gathering forms to villages through localised NGOs.

- The central NGO signed arrangements with these localized NGOs. The latter identified local farmers who were trained and served as "animateurs". They explained the project to kin and neighbors, recruited tree planters, organized pre-planting meetings, coordinated the delivery of the seedlings, and gathered follow up survival data on every plot on which trees had been planted.

Results

1) The project hesitatingly agreed to plant three million trees on peasant land between September of 1981 and September of 1985--hestitatingly, because it was all founded on untested anthropological predictions about peasant tree planting behavior and institutional behavior. The hypotheses have been totally validated: as of this date (May '85) already more than 15 million trees have been planted by some 30,000 Haitian peasants. Other approaches to tree planting run into the problem that peasants will not plant trees, which then die in nurseries. The main problem of the Haiti Agroforestry Project, in contrast, is that it is difficult to produce seedlings fast enough to satisfy the demand that has been created. Cases were reported in which peasants who were not recipients of project trees actually uprooted and stole samples of newly planted seedlings

from other villagers' plots. The project has managed to touch an economic nerve in Haitian peasant villages.

2) Though survival rates of the trees have not been precisely calculated yet, it is clear that in moist areas survival rates are high. In arid areas they are low. What is important is that mortality is due, not to human carelessness (e.g. livestock depredation, poor care) but to ecological conditions. Peasants are according to the trees the same degree of protection which they accord to their crops, which was a central project goal.

3) Final planting decisions were left to the peasants. The vast majority of trees have *not* been planted on marginal land, but rather border planted or intercropped on plots where the peasants are simultaneously growing crops. The reasons underlying this behavior lie less in the labor-saving hypothesis of some observers ("the peasant achieves two crops with one ground-preparation") than in land tenure dynamics. Agriculturally marginal land is generally left by kin groups in undivided blocks for common grazing purposes. Only agriculturally productive land is subdivided into individually controlled plots. This means that the tenure relations prevailing on marginal land are such as to impede clear tree-tenure rights. No matter who plants a tree, any kinsman with collective rights in the land can cut the tree down. Thus land tenure dynamics create a situation in which land whose commercial value could be increased by trees (marginal land) will not be planted in trees because of the tree-tenure insecurity that would affect such land.

4) The impact of this project on the behavior of institutions varies according to the type of institution. The participating NGOs have changed their approach to tree planting in a manner that is probably irreversible. It is unlikely that they will revert to the less effective models that prevailed before this project. In addition a Swiss donor organization (Helvetas) made a major financial contribution to the project after viewing its operation. Such international support opens the possibility that applicable features of the model may spread to other world regions.

5) Public sector institutions, in contrast, cling to old patterns. The Haitian government has given no indications that its own tree planting procedures will be influenced by the successes of the Agroforestry Project. The only response forthcoming from the Ministry appears to be continued irritation that the funding was channeled through pri-

vate organizations, thus violating its sovereign right to squander donor funds. USAID, the major funder of the Agroforestry Project, also gives evidence of returning to the old ways. Discussions are under way for a multimillion dollar "Watershed Management Project," to be funneled through the Haitian government, using organizational and operational tree planting models that have already failed spectacularly in Haiti. The behavior of these two governmental institutions is to be attributed less to institutional stupidity than to the possible attachment of some decision makers to agendas which have little to do with tree output.

6) Former accusations of land tenure insecurity as the major cause of tree planting failure in Haiti have been rendered untenable. A land tenure system, rather, with its internal heterogeneity, can be seen as a "menu". The goal of planning is to target that subset of plots which can serve as the stage for unleashing the same economic energy toward wood cultivation that human populations have for some 12 millennia channeled in food cultivation. If planners fail to discover the potential linkages, it is their fault, not that of the "land tenure system". There may be regions where such wood planting is truly hindered by land tenure. But I suspect that many more cases of failure are due more to the mediocrity of the planning and management process than to inadequacies or insecurities of the land tenure system.

Agroforestry Issues in Latin America

Marion R. Brown

Land Tenure Center, University of Wisconsin-Madison

The predominant feature of development in rural Latin America in the post-World War II era has been rapid modernization and mechanization of production on large holdings.

This trend has been accompanied by dramatic shifts in land use and tenurial and labor arrangements. Food crops for domestic consumption have been displaced by crops for export. Permanent resident workers, share-tenants, renters, and small holders have been pushed off the best agricultural lands in fertile, often irrigated, valleys onto hillsides and into forested areas.

These displaced peasants are not swidden farmers steeped by tradition in ecologically relatively sound practices of shifting cultivation --they are newly landless farmers accustomed to deep alluvial soils and flat or gently rolling landscapes. They have not fared well on steep slopes or in rain forests nor have the resources they utilize to eke out a subsistence. The result is a crisis of shifting agriculture and a human and ecological disaster of major proportions.

The hopes of agrarian reformers in this context have not been realized. Serious efforts to redistribute fertile agricultural land have been thwarted or reversed in country after country. Resettlement schemes--often implemented as substitutes for redistributive reforms--have aggravated the problem. More often than not, they have only proved that lands thought to be unoccupied were, in fact, already being quite rationally utilized by swidden producers and were usually too fragile for continuous cultivation.

In most cases, land reforms did not include much by way of land-use planning, much less agroforestry as it is conceived today. In Chile, some large privately owned forests were expropriated and there was a serious but marginally successful effort to establish agricultural and forestry enterprises. In El Salvador, land-to-the-tiller reforms may have made matters worse by interrupting the pattern of shifting cultivation practiced by renters of small hillside plots. This is a matter of debate, with some arguing that landowners kept their renters on-the-move to prevent them from becoming too attached to any given plot of ground, and that this system did not include any significant fallow period. Others contend that renters moved from plot to plot each year in a systematic five-to-six year rotation which allowed native leguminous, woody plants to reestablish themselves between cultivation cycles.

In short, three decades of rather rapid changes in land use and tenurial arrangements have not brought about a more harmonious relationship between field and forest nor established sound bases for modern agroforestry in Latin America.

The basic tenure problem remains one of a highly skewed distribution of land which forces overuse or inappropriate use of marginal and fragile lands and destruction of trees and other forest products.

The following are brief and admittedly oversimplified responses to each of the questions put to us in the invitations we received from John Raintree to prepare position papers for the workshop.

1) *What are the main tenure problems and how are they related to the adoption of agroforestry innovations?*

The big issue continues to be concentration of landownership. *Minifundistas*, renters, sharecroppers, and shifting cultivators operate on the edge of survival where immediate food and fuel needs outweigh long-term needs and opportunities, including those of agroforestry.

2) *What agroforestry systems show potential for solving or mitigating existing tenure problems?*

It seems highly likely that smallholder agriculture in Latin America could benefit from additional emphasis on tree crops that provide food, fuel, shelter, protection from erosion, or soil improvement through nitrogen fixation. Fuelwood as a cash crop for smallholders (as in the Haiti project) also appears to have potential, though middle-income farmers could be expected to benefit more than those on very small parcels.

3) *How do these considerations relate to the problems of disadvantaged groups (e.g., women, the landless, etc.)?*

This question is closely related to the first. The basic tenure problem is access--and equity of access--for the rural poor. Modernization of the existing structure has not favored these groups; rather, it has worsened their lot. Where agroforestry is concerned, a major mistake has been to think of unused or underutilized areas (often rain forests, semi-arid regions and hillsides) as agricultural frontiers where disadvantaged groups could be resettled to relieve pressure on existing farmlands. It would, in my opinion, be a step in the right direction to simply abandon the idea of relegating the disadvantaged to "reservations" in fragile ecosystems. Where settlement schemes are considered, much more attention needs to be paid to the technologies of agroforestry and to peasant capabilities and penchants for using these technologies in ways that are economically sensible and ecologically sound.

5) *What research is needed to clarify particular tenure issues in agroforestry?*

I believe a priority research focus should be the apparent crisis in shifting cultivation in Latin America. What is the extent of the problem? Why are relatively sound swidden systems breaking down? More specifically, what have been the effects of population growth, displacement and resettlement, and land reforms of various kinds (e.g., the land-to-the-tiller reform in El Salvador which may have "stopped" the rotation cycle and created new problems for both peasants and the ecology)?

6) *How do the above considerations relate to the planning and implementation of research and development projects (e.g., methodological and institutional issues)?*

There are many facets to this question. I will respond to only one --the urgent need for more effective interdisciplinary efforts involving physical and social scientists. There are encouraging signs here: political scientists and rural sociologists working with departments of forestry at major universities; ICRAF's emphasis on multidisciplinary activities; the presence of foresters, sociologists, economists, and anthropologists at this conference. Still, I often hear definitions or concepts expressed by one group that appear to ignore the concerns of the other. I can offer nothing by way of a solution except encouragement to us all to keep trying to be more genuinely communicative in our inter- and multidisciplinary initiatives.

Part 3

Working Group Reports and Plenary Discussions

Report of the Regional Working Group on Africa

Chair: H.W.O. Okoth-Ogendo
Rapporteur: David Brokensha

The wide ranging discussions of this rather large and lively working group are summarised here under five major headings

Customary vs. statutory law

Conflicts between tenure systems are common in Africa, particularly between local customary and national statutory tenure regulations. Where two systems exist together, each with a fair degree of credibility, uncertainty is created and customary rights may be wiped out.

Statutory law in Kenya, Zambia and Malawi stresses private property. In Zaire since 1973, the government is the only recognized "landlord," although traditional rights are accorded recognition through "le grand principe" of traditional land law. Although this approach to regularization of tenure is an interesting one, doubts were expressed about its applicability to countries with a very heterogeneous population, where the rights of minority groups might suffer if the dominant group is all-powerful in politics.

Manipulation of laws for private gain is a common problem in Africa. This is most easily accomplished when there are two tenure systems in operation. Community development projects may be "hijacked" by local elites for their own benefit, as in Kenya where young educated Maasai have registered and secured exclusive title to land improved by projects. These kinds of developments are not the exclusive province of local elites, however, as in northern Ghana where "strangers" from outside the local community have used registration to take control of land away from the locals, who had unwittingly assumed the continuation of traditional rights. Similar examples of alienation of local land rights may occur where government ownership of all unregistered land is a legacy of colonial governments, or where large areas of land are taken for "national interests," e.g. large mechanized farms in the Sudan.

As regards private property, national legislation tends to emphasize soil and land, whereas customary law takes more account of trees. An important research question is: What is the degree of overlap or difference between tree tenure and land tenure in particular areas? The group also noted the importance of distinguishing between "individualization" (a widespread trend in Africa toward individual control of land *use*) and "ownership" (usually including rights of disposal).

Governments in Africa often regulate the use of "national resources" in ways which impact on agroforestry development. For instance, certain tree species may be protected, so that it is prohibited to cut even those trees which occur on a farmer's own land. In many cases this goes so far as to prevent farmer's from harvesting trees which they themselves have planted. Regulations designed to protect forest reserves often prohibit entry altogether. In other cases, entry is allowed for specific purposes (e.g. grazing, collection of fuelwood, fruit or honey) at different stages of forest development, so that local people have certain legally stipulated rights. Given the need to renew the national resource base, an important research question is: What are the rewards and disincentives for tree planting?

Timber concessions are another important aspect of the land use scene in Africa. Governments may treat trees like minerals, allowing concession holders to remove all trees of a certain species from the land, even on individual holdings. In reaction to this, farmers in Ghana have been known to cut down valuable trees rather than risk the damage caused by concessionaires in getting access to the trees.

In regard to the implications of all this for agroforestry, a major need is for an increased awareness of the complexity of the situations in which agroforestry projects take place. The group felt that where tenure conflicts exist, the attempt to change land tenure systems is not necessarily the best approach. It may be easier to achieve development goals (including equitable distribution of benefits) by changing market conditions instead.

Major tenure trends in Africa

Sharecropping is becoming a more prevalent feature of land use in Africa. A recent eight country survey showed that this was a major innovation in all of the countries. An important research question is: How does sharecropping affect the planting of trees? If tree planting by the land owner is integral to the process by which land is brought into production (e.g. as is typically the case with cocoa), then there is no particular problem. However, if tenants want to plant trees and the landowner sees this as a threat to the security of his tenure over the land, then serious problems may arise, particularly where the planting of trees establishes the basis in customary law for a claim to land. One way of handling this might be to persuade the landowner to plant the trees, and then allow the tenant to take a contractual share of the harvest in return for his labor inputs.

African farmers are increasingly experimenting with new forms of land use in order to cope with population pressure on traditional land use systems. The group felt that local solution-finding should be encouraged and noted, in this connection, that customary tenure law is not rigid, as it is often portrayed, but flexible and innovative. Local people should be helped to find ways of accommodating traditional tenure systems to new and more productive patterns of land use. The group noted that existing land tenure systems can provide a "menu" from which the elements of a workable model may be taken.

As regards the appropriate unit for agroforestry projects, the working group noted that the emphasis tends to be on a farm level approach (i.e. individualized tree planting efforts on private farms), but that many situations still demand some form of group action. It was pointed out that there is a strong social dimension to land use in many African societies, which often requires an individual to obtain group approval before planting trees. Also, if agroforestry interventions are undertaken on a large scale for ecological reasons (e.g. watershed protection), then group action may be an inescapable requirement. In some cases there may be a need to institutionalize group sanctions against non-joiners or community members who hinder the tree planting efforts of others. More generally, the group consensus was that different units are required for different areas.

Special problems

Among the categories of land users that present special problems to agroforestry development and require especially innovative solutions the group singled out three for special mention: pastoralists, the landless and near-landless, and women.

Very little systematic knowledge exists concerning tree use and tree planting behavior among pastoralists. Among the critical questions and concerns to be addressed by research are: What are the rules regarding tree rights? What are the rights of individuals *vs.* groups? For example, in parts of Sudan, the family has the exclusive right to shake fodder pods from certain trees, although others may allow their animals to eat the pods which have fallen naturally to the ground. In Maasailand, it is common for group ranches to allow herders from neighboring group ranches to graze on the outer fringes of one ranch's territory. However, during the recent drought, the membership of at least one group ranch with an abundance of *Acacia tortilis* trees on one of their outer borders activated exclusive tree shaking rights, restricting non-members to the use of naturally fallen pods. This kind of measured response to the drought suggests more sophisticated management of tree resources than is commonly ascribed to pastoral groups.

How can trees used periodically by nomadic pastoralists be protected when the pastoralists are not present? Newly arrived immigrants into pastoral land may be ignorant of, or simply ignore, the unspoken rules that govern relationships between existing groups, and they may tend to be more opportunistic and aggressive in the appropriation of common resources. What are the attitudes of government? Often there is strong pressure to encourage sedentarization of nomadic or semi-nomadic pastoralists, which may not be the best solution from an ecological point of view. Also, increasing population in well-watered pastoral areas is putting further pressure on trees and grazing resources. Does this mean that pastoralists will become more interested in planting and caring for trees?

As regards the landless and near-landless, it was noted that inequality of land distribution is a major problem in some parts of Africa. In forested areas where agricultural land scarcity is severe, foresters charged with the protection of reserved forests are confronted with increasingly insistent problems of forest encroachment. Taungya schemes may offer a partial solution, but only if they are specifically designed to provide long term, equitable land use opportunities to the farmers. In general, although conventional block planting approaches to tree planting offer inherently limited opportunities to smallholder farmers, agroforestry intercropping approaches may offer significantly greater scope for tree planting on small farms, without reduction in crop yields.

Several problems specific to women producers were noted as having an important bearing on agroforestry development. In Africa as a whole, some 30 percent of rural households are headed by women, who typically have fewer rights in land and trees than men. Inheritance patterns tend to favor men; and land tenure changes associated with land registration and adjudication usually have a male bias, thus increasing the insecurity of female producers. The work load and division of labor, moreover, is different for women and men, and women tend to have different responsibilities and rights with regard to trees. Access to cash crops, including trees, tends to be a male dominated area, but some woody plants are regarded as being in the female domain (e.g. *Sesbania sesban*, a valuable agroforestry tree in the Kakamega District of Kenya). The general conclusion was that women usually have diminished rights in land and trees and, unless specific measures are taken to ensure equitable opportunities for women, they are likely to be excluded from full participation in agroforestry projects. Even where tree planting projects depend primarily on female labor, the benefits are not likely to be enjoyed by women. To the extent that agroforestry can help increase food production, it is in everyone's interest to include women, since they are the main food producers in Africa.

Problems of awareness and information

The group singled out three categories of actors on the agroforestry scene which are in need of special awareness raising efforts: government departments, donors, and the farmers themselves.

One positive change in the orientation of government foresters is already apparent; namely, the greater tendency to look at forests from the viewpoint of the people and to allow limited use rights in reserved forests rather than simply excluding everyone. Although recent changes in the training of African foresters have put more emphasis on sociological aspects of the job, training institutions have yet to reorganize sufficiently to meet the new training demands. It is not enough to have a token sociologist on the staff.

Agroforestry is now quite popular with donors, but they are constrained in the effectiveness of their response by such factors as the need to move large amounts of money quickly and the lack of time allocated to the planning of sound agroforestry projects. As yet in agroforestry there are few quick fixes, and time is often needed to develop the research basis for large scale agroforestry development schemes. Much of the requisite information is held by the local people, and project planners are rarely able to take the time to get an adequate appreciation of local knowledge and preferences. With regard to tenurial aspects of agroforestry in particular, the state of formal knowledge is woefully inadequate. Funding for more research is urgently needed.

While acknowledging the relevance and depth of indigenous knowledge, the working group also felt that most farmers are still largely ignorant of the range of options open to them in agroforestry and emphasized the importance of information dissemination and demonstration.

Research priorities

Some research should be project specific, other should address more general issues. The working group also felt that in research there was a tendency to overlook local researchers and place too much emphasis on expatriate "experts." To offset this tendency the group recommended that research projects should make an effort to include local researchers. The following questions and concerns were identified as priority areas for research:

1) What is the appropriate scale of social organization for different forms of agroforestry? This question should be considered in light of the constraints on smallholder farming, e.g. small size, fragmentation of holdings, labor shortage, etc.

2) Selection of species for agroforestry should be chosen in relation to tenurial systems and other social factors as well as technical factors.

3) The relationship between tree planting and tenure is a major research question. When does tenure obstruct, and when does it encourage tree planting? What are the inheritance patterns for tree rights?

4) Many recent and dramatic changes in land tenure have led to insecurity. What external factors (e.g. population increase, changes in the national legal code, market forces, etc.) lead to new land use and tenurial patterns (e.g. sharecropping)? Which of these are most directly related to tenure insecurity?

5) How can the landless and near-landless share in the benefits of new agroforestry systems? Is it feasible for them to receive limited rights in public lands for agroforestry production?

6) Little is known about the whole question of tree tenure among pastoralists. This should receive more attention.

7) More research is needed on the tenure rights of women in regard to both land and trees.

8) What are the most effective institutional arrangements for the promotion of agroforestry? Which government agencies should be involved (agriculture, forestry, natural resources, livestock, energy, environment, etc.)?

Report of the Regional Working Group on Asia

Chair: Sediono Tjondronegoro
Rapporteur: Michael Dove

Tenure systems in the Asia region tend to be highly evolved, quite varied and, in their interaction with agroforestry, extremely complex. While the deliberations of the working group reflect this complexity, some general comparative perspectives did emerge.

Comparative perspectives

Despite widely varying colonial and post-colonial experiences, there are broad similarities throughout the region in: 1) the tendency of governments to regard all forested land as State land, and 2) the related tendency of government foresters and forest-dwelling peoples to interact in an adversarial fashion. Among the most important dissimilarities within the region, the group noted that the issue of seasonal grazing rights on temporarily uncultivated agricultural land, while important in Central and South Asia, is rare or non-existence in Southeast Asia.

In comparing the Asia region to Africa and Latin America, the group noted that government concern about forest dwelling tribes is a major issue in Asia and Latin America, but not in Africa. It is also noteworthy that among the cognatic, egalitarian tribal societies of the Southeast Asia region, there is little or none of the gender-based division of property, labor and the fruits of labor that is so prominent in the African and South Asian accounts. This is an interesting observation in that it demonstrates one of the central tenets of feminist critiques of sociobiological explanations of gender inequality; namely, that gender differentiation and inequality are not universal characteristics of traditional societies, but rather a function of particular historical, economic, political and social circumstances.

One of the most striking regional differences is the sharp dichotomy between stratified peasant societies practicing intensive agriculture in the lowlands of Asia and the egalitarian tribal societies practicing shifting cultivation in the uplands, which appears to be absent as a recurring theme in most of the African and Latin American material. The explanation for this may have to do with the particular character of the Asian environment and/or the historical development of the characteristic Asian State.

In general, the role of the State appears to be a more important factor in the planning and implementation of agroforestry programs in Asia than in Africa and even Latin America, although in the latter case the contrast is less sharp. This is reflected in the unanimous opinion within the working group that any attempt to bypass Asian governments in the planning and implementation of agroforestry programs is completely impractical. The source of this difference appears to lie in the extremely centralized and hierarchical character of both historical and contemporary Asian States.

Main tenure issues

State attitude toward peasant rights

With some exceptions, such as the recognition of peasant rights to pasture in state forests by the Nepalese government, the governments of this region recognize no peasant rights to forested lands. Rights to non-forested agricultural land are often perceived by governments as communal, even where individual rights prevail or where only residual rights are communally held.

Communal land rights

Where communal rights exist, they are often only of a residual nature; just as often no communally held rights exist. Where such rights exist, it is important to recognize that their function is not always the same. In egalitarian tribal hill regions, communal rights may be held by the group as a whole; whereas in stratified lowland peasant areas, they may be used only by the elites. Where present, communal rights may complicate tenurial issues and incentives for agroforestry.

Land fragmentation and concentration

Given relatively unrestrictive systems of inheritance common in the region, increasing population density creates a tendency toward greater fragmentation of land holdings. At the same time, the trend toward increasing capitalization of agriculture, along with other political economic factors, is leading toward greater concentration of land holdings in some areas. The former trend raises questions of productivity, the latter questions of equity. Assessments of both productivity and equity are relative to different criteria used by different groups. For example, productivity might be a measure in terms of: 1) returns per unit of land (as is common among irrigated rice farmers), 2) returns per unit of labor (as is common among swidden rice farmers), 3) carrying capacity (a measure used by ecologists and demographers), or 4) extractable surplus (the measure commonly used by governments).

Alleviation of tenure problems through change in legal systems

The group felt it was important to integrate customary law, especially that involving traditional rights to forested lands, into the framework of national legal systems. Viewpoints on the primacy of one or the other system often seem irreconcilably different. The view of officials that peasants have been illegally occupying state lands without realizing it has a mirror image in the peasant view that the state has been illegally appropriating private peasant lands without realizing it.

Where formal laws have been changed, positive impacts on agroforestry can be noted. One example comes from the Philippines, where a forest-dwelling tribal group induced the government to recognize their stewardship of a part of the forest, and then developed a very successful system of agroforestry on their own initiative. In other examples, from Nepal, the granting of 99 year leases on state forest land gave erstwhile shifting cultivators sufficient incentive to actively participate in the successful rehabilitation of the land using agroforestry techniques.

Land reform and redistribution may be a desirable component of some agroforestry projects, but care must be taken to try to predict all of the possible consequences. In one case, again from Nepal, a land reform program stipulated that non-owners who worked a given piece of land for three years thereby acquired title to it. The apparently unanticipated consequence of this program was that landowners chose to leave fallow all land they could not work themselves. This intervention might have resulted in the extensive planting of low-labor tree crops, but instead it resulted in the rapid deterioration of terraces on unworked land.

The working group concluded this discussion with two suggestions: First, that a special national judicial agency should be established to deal with tenure problems arising from agroforestry development. And second, that government should stop using pejorative classifications (like "minority group" or "foreign tribes") in their dealings with forest dwelling nationals.

The appropriate unit and focus for agroforestry programs

The group found it difficult to reach a consensus on whether agroforestry programs should focus on individual farmers and private land or on groups and communal land. It is likely that different approaches will be needed for different groups. Thus, a communal land development scheme might appeal to landless laborers, but it would not be likely to appeal to farmers who already owned land. However, the group noted that lack of incentive for private participation is a problem that has plagued communal schemes and for which no really satisfactory solution has yet been found.

It was also difficult for the group to decide whether the real focus of an agroforestry program should be defined socially (e.g. a village) or ecologically (e.g. a watershed). In the end, the consensus emerged that for agroforestry purposes the watershed would probably be a more appropriate unit. The group noted that villages could also be studied using ecological concepts and went on to add the following caveats to their endorsement of the watershed as a priority unit for agroforestry: First, the human ecology of a watershed can be profoundly affected by regional (e.g. urban) variables beyond its boundaries. Second, a large watershed may contain so much variation as to require not one, but several agroforestry development plans. Finally, the use of ecological criteria to define the planning unit may result in more attention being paid to ecological than socioeconomic priorities, unless special care is taken to prevent this.

Program initiatives and planning

Experience from Indonesia was cited which suggests that government funded land development schemes tend to draw the attention of predatory elites and should be avoided as far as possible.

The group agreed that the most desirable way to plan and implement agroforestry programs is with a bottom-up approach, using peasant organizations. However, a problem with this approach is that, while peasant organizations already exist in many of the countries in the Asia region, most of them have in effect been created by the State and may not faithfully represent the true interests of the peasants. There is, thus, a problem in thinking of these organizations as a basis for a truly bottom-up approach.

The working group also felt that, ideally, it would be desirable to plan and implement agroforestry programs through NGOs. However, the reality in most of the countries of the region is that this may be a politically infeasible and unsustainable approach. On the other hand, the alternative of working solely through governments cannot be regarded as a recipe for the future, because government priorities are likely to differ from those of the peasants. The group concluded that the only practical alternative is to attempt to work *both* through central governments and through NGO's, with the former providing legitimacy and the latter working to ensure it. The Philippine experience suggests that for the implementation of projects at the pilot stage, advisory or working groups of mixed composition may be preferable to either governments or NGOs alone.

Research priorities

Evaluation studies

Much of the literature on agroforestry is normative--describing programs, urging their implementation, etc.--as opposed to evaluative --analyzing when, where and why particular agroforestry programs have succeeded or failed. No doubt this is partly explained by the relative scarcity of mature agroforestry projects to evaluate. Agroforestry represents a new field of organized project activity, and one that takes comparatively longer to obtain results. The validity of this excuse diminishes with every passing year, however, and it is beginning to be time to activate a more evaluative approach to research. Such research should address the following questions, among others:

What, if any, were the unintended consequences of particular agroforestry technologies and development programs? What has been the impact on the local ecology? What has been the impact on the economy of the local participants as well as on the regional economy? (The answer to this question should include an assessment of the costs as well as the benefits, and comparison with traditional land use systems and alternative technologies.) Of particular relevance to Asia, where shifting cultivation is one of the most appropriate targets for agroforestry programs: Which components of existing swidden systems are most and least responsive to development by agroforestry techniques? Finally, based on the above, what are the characteristics of an appropriate agroforestry technology for a particular set of ecological, economic, cultural and political circumstances?

Baseline studies

Given the recency of the scientific "discovery" of agroforestry, as compared to its long history as an indigenous land use practice in many, particularly Asian, populations, it is reasonable to expect that many of the traditional agroforestry systems are far richer and more sophisticated than many of the systems that researchers will be able to develop in the near future. It is important, therefore, to study traditional agroforestry systems and to make full use of indigenous agroforestry experience in the planning of contemporary agroforestry projects.

For research on these indigenous systems of empirical knowledge, the methods of ethnoscience (ethnobotany, etc.) are particularly relevant, especially as regards indigenous knowledge of and classification schemes for trees, non-tree plants, etc. and for various types of forests, farms, gardens, and other forms of land use. This knowledge will be particularly fruitful if it can be related to local perspectives on the production motivations, opportunities and limitations of socially defined categories of land users (e.g. women).

Policy studies

What is the relationship between colonial forestry policy, contemporary government forestry policy, and agoforestry policy? Are there important distinctions between the explicit (manifest) and implicit (hidden) functions of contemporary forest policy in regard to resource management and social control? What are the implications for the planning and implementation of agroforestry programs?

Development studies

Where do agroforestry programs fit in the overall scheme of national development? How does the role for agroforestry vary between and within regions or other development units? For example, within the same watershed the main role for agroforestry in the uplands may be in controlling erosion, whereas in the lowlands it may be in providing opportunities for the landless. What is the appropriate unit for agroforestry projects? What kind of institutional support is necessary for the success of agroforestry programs? Some of these questions imply a solution in search of a problem, when in fact we all agree that development projects should start with an analysis of the problems and only then proceed to the identification of solutions.

Training

What kinds of training are needed for the planners, implementors, and participants in agroforestry programs?

Report of the Regional Working Group on Latin America

Chair: Rafael Baraona
Rapporteur: Dianne Rocheleau

Client groups for agroforestry

The working group first tried to define the range of land use and tenure systems operated by different client groups at places in the Latin American region where agroforestry is either incipient or already in practice. Six different client groups were identified: 1) forest dwellers who manage and make intensive use of trees, 2) shifting cultivators, 3) farmers who use a combination of on-farm resources and open access to off-farm resources (e.g. fodder, fuelwood, timber, food, etc.), 4) farmers and farm laborers who make use of concessionary forest rights or who trade in forest products, 5) farmers with diverse multipurpose trees in orchards, homegardens or some kind of dispersed tree arrangement throughout the farm, where the trees are a substantial component of a mixed subsistence/commercial farm economy, and 6) farmers specializing in trees as cash crops.

Land use systems

In analyzing tenure relationships the group felt it important to identify the relevant units of social organization and levels of spatial integration in the operational environment of different client groups (e.g. integration of trees, crops, animals, farmers and herders at the plot, farm, watershed, community, ecosystem, and/or regional levels). For example, a special kind of tenure problem arises when a nearby forest is regarded by the farmer as an operational part of his/her "farming system." In general, different tenure issues arise at different levels of social and spatial integration of land use, and the group felt that this was a useful way of looking at the complexity of tenure relations.

As regards the diversity of land use systems involved in agroforestry, the group felt it was important to use a theoretical perspective that integrates human ecology (e.g. population pressure and land use intensification) with critical aspects of political economy. Other land use variables in agroforestry which have important tenure implications are: the size and quality of land holdings in relation to formal and informal aspects of tenure security and the relative importance of various forms of shared land use with different tenure arrangements (e.g. commons, public land, the holdings of farmer associations, etc.).

The group noted that in Latin America there is a different range of land-sharing arrangements than is found in Africa and Asia, particularly as regards the strength of traditional forms of group tenure.

The group emphasized the special significance of homegardens as an agroforestry system with relatively high tenure security, which can be intensively developed to play important nutritional and cash income roles in the household economy. The importance of preserving or substituting for open access areas was also noted, particularly in connection with the maintenance of browse resources in systems of short term transhumance in mountain environments.

Participation and equity concerns

The group emphasized the importance of letting people define their own development objectives and future directions as regards land and tree use. In this connection the importance of understanding relevant aspects of indigenous science and the local social organization of production was underscored. The need for responsive technical assistance was also stressed, particularly in providing relevant market analyses and forecasts to help the rural poor make profitable innovations that work for them. In this connection, the problem of "too profitable" projects or enterprises being captured by local elites, large commercial interests, or government was noted.

The need to clarify and strengthen women's legal access to land, credit and technical assistance was highlighted in relation to land tenure reform, settlement projects, and agroforestry programs. Special emphasis should be given to female-managed households and the tenure rights of women in common law unions. The group also stressed the importance of understanding the sexual division of labor, resource utilization and decision-making in households, in order to ensure that all members of the household are able to participate in the planning of and benefits from changes in land and tree use patterns.

Institutional and administrative issues

Among the institutional issues discussed by the working group and flagged for further attention were: government tree and forest tenure *vs.* individual rights, which is a particularly crucial concern for agroforestry in Central America and the Caribbean; government *vs.* NGO administration of agroforestry programs; forest policy *vs.* agrarian reform policy as areas of policy critical to agroforestry; and the need to improve forest tenure for indigenous groups.

Research priorities

The working group identified several priority research needs and questions under three headings: legal and institutional, land use and management, and participation and organization.

Legal and institutional

1) There is a need for a comparative review of the legal and institutional situation in each country in regard to tenure regulations affecting the use and planting of trees, with particular attention to development conflicts between foresters and farmers. Such a study could start with an effort to define what constitutes a "tree" and a "forest" in the eyes of the law and its enforcement agencies.

2) Also needed is a comparative study of the approaches used by different development institutions (international agencies, government institutions, and NGOs) to agroforestry, tenure and participation questions, with a view toward learning from past successes and failures.

3) To what extent is formal tenure insecurity actually an obstacle to agroforestry development? What effect does the formalization of tenure have on clients and on agroforestry systems? What are the mechanisms that provide for security in informal tenure systems (both local and national)?

Land use and management

4) What are the existing and incipient varieties of agroforestry in the region and what are the land and tree tenure situations in each case?

5) What and where are the crisis points and trouble spots in relation to shifting cultivation, deforestation, over-exploitation of browse resources, and rapid shifts in access to land and tree resources? Who and where are the clients for immediate action on agroforestry and tenure?

6) How can the use rights in commons, absentee-owned, or public lands now used as open access resources by communities be strengthened, legalized and regulated for sustained productivity? This subject might be analyzed from a variety of perspectives (e.g. by activity, product or tree; as a block *vs.* a dispersed resource; by individual, group, community or government institution). Solutions will

vary with the situation, but what are some of the most promising options?

7) How can the browse resources of the commons best be managed? This research should seek to identify the most promising combinations of technology, management and legal measures to solve the problems of the dry season fodder gap, degradation of resources, and distribution of benefits and responsibilities in particular situations.

Participation and organization

8) What are the current and forseeable conditions of access to markets? What are the current and near future markets for agroforestry products? What are the potentials for locally beneficial processing industries?

9) What effect do tenurial and organizational history have on a group's preference for and capability to manage different agroforestry/tenure combinations? Case histories are needed of relations between agroforestry and tenure and their adaptation to changing conditions of environment, land use and political economy.

10) What constitutes a "tree," a "forest" and an "agroforestry system" in the eyes of the local people? How does this affect their motivation to participate in agroforestry projects?

11) Comparative studies are needed of different project approaches to agroforestry, tenure relations and participation in similar situations (this is similar to number 2 above, but with the emphasis on controlled situational comparisons). What promising models exist or could be developed for action research emphasizing on-going evaluation and modification of agroforestry innovations by local participants?

Plenary Discussion on Regional Issues

Chair: Marion Brown
Rapporteur: Luzviminda Cornista

This was quite a wide ranging discussion, but several major issues emerged from the discussion of the regional reports and are summarized below.

On broad similarities and contrasts between regions

It appears that both Latin America and Asia have gone through similar historical processes of land fragmentation, land concentration, tenancy, and State control of land. There are numerous lessons to be gleaned from comparative analysis of these histories. Africa, which emerged from the colonial period only quite recently, shows signs of embarking on the same path. The hope was expressed that Africa could learn from the lessons of the other two continents and avoid the problems associated with this course of development.

On diversity and variations between and within regions

Considering the high degree of inter- and intra-regional diversity in regard in tenure systems, it was felt to be hazardous to make too many generalizations about regional similarities and differences. The general consensus was that agroforestry development should start from the assumption of heterogeneity and pay close attention to the case specific nature of the tenure situations in which agroforestry is to be operationalized.

On government involvement in agroforestry

Two schools of thought were expressed on the role of government in agroforestry development. The first argued that government involvement in agroforestry programs is not functional and should be avoided whenever possible. Working directly with NGOs was felt by this group to be a more productive route to agroforestry development. The second school of thought argued the need for government involvement in agroforestry. It was pointed out that governments exert a pervasive influence in most countries and that to ignore them would mean either that agroforestry programs would not be launched or that they would not be sustained. The inevitability of working through governments was the general consensus of the workshop.

On the role of forestry departments and their relationship with other government agencies

The need to look beyond the forestry department of the government as the lead agency in agroforestry initiatives was emphasized. The fact that agroforestry, by its very nature, includes agricultural and livestock production means that the agencies responsible for these components must also be involved in agroforestry programs. An enduring problem, however, is the question of coordination and integration in program planning and implementation at both national and local levels. Thus, a relevant area of research is on institutional behavior and relationships, with a view toward uncovering the problems and constraints impeding an integrated approach to agroforestry.

The need to reorient the role of the forester from forest guard to extension agent was also underscored. This implies three considerations: first, the retraining of foresters for the necessary technical and social skills; second, changing the composition of the staff of forestry departments to incude anthropologists and sociologists; and third, inclusion of more social science subjects in forestry courses.

On the role of NGOs

The role of NGOs in agroforestry development was recognized as important, although it was stressed that not all NGOs are equally competent to implement agroforestry programs. That being the case, evaluative studies need to be undertaken to identify potential roles for particular NGOs in a) making government work better, b) managing programs, and c) helping to develop institutions for processing and marketing of agroforestry products.

On donor agencies

It was pointed out that donor agencies can become constraints to agroforestry. For instance, attempts by governments to adopt a multiagency approach in agroforestry are sometimes thwarted because such an approach may not fit the framework of the donor agencies. In other cases, grants and loans may do more harm than good because they are not well thought out. To make investments more meaningful, quality rather than quantity of investment should be the guiding principle.

On the operational unit of agroforestry programs

The question of whether to use an ecological unit, a social unit, or an administrative unit in the implementation of agroforestry programs was discussed. In determining the focal unit, the importance of considering both the physical and the human ecological aspects was emphasized. Consequently, a typology of land use systems is necessary which incorporates both demographic and political economic dimensions of land use.

On peasant organizations and land groups

Although it was observed that most peasant organizations (particularly in Asia) are government initiated, their role in articulating the interests of the peasants and in sustaining development programs should not be overlooked. Furthermore, the existence of other types of social groupings, outside of the traditionally labelled social units such as family, lineage, tribe, etc., should also not be overlooked. For instance, there are "land groups" whose reason for existence is their common history in a political struggle. Such groups could also be appropriate social units for agroforestry.

On other research areas and data needs

It was pointed out that researchers on tenure and agroforestry have not given sufficient attention to the analysis of policies and laws that provide the legal framework for development programs in general and agroforestry in particular. Moreover, an examination of these policies and laws from an historical perspective would provide a more holistic understanding of the context within which tenurial considerations in agroforestry should be viewed.

Finally, the need for good regional information on basic demographic, land use and tenure data was stressed.

Report of the Working Group on Technology-Tenure Interactions

Chair: Percy Sajise
Rapporteur: William Bentley

In considering the interaction between agroforestry technologies and tenure systems, the working group began with the observation that the interaction can go in either of two directions: 1) existing tenure relations can have an impact on the adoption and distribution of benefits from new agroforestry technologies, and 2) new agroforestry technologies have the potential to modify the existing structure of tenure relations within an area. For each type of interaction, the impacts can be both positive and negative.

Much of the ensuing discussion centered on how to maximize the positive and minimize the negative impacts of technology-tenure interactions. In an effort to produce some general guidelines, several different approaches to the construction of a technology-tenure interaction matrix were considered. Although this was thought to be a useful exercise in the context of a specific agroforestry project, a general framework proved difficult to achieve in the time allotted to the working group, due to the great variety of agroforestry technologies and tenure systems and the complexity of the numerous site-specific interactions which could potentially occur. The exercise, nevertheless, was fruitful in uncovering a number of important considerations and in highlighting the points of general agreement which are summarized in this report.

It was recognized that technology-tenure relationships are likely to vary considerably between ecological zones, particularly in response to differences in soil fertility and moisture stress. Dynamic relationships between population pressure and land use patterns within a particular environment provide one effective starting point for an examination of technology-tenure interactions. Another approach is to identify the generally desired attributes of technology and then to assess how these interact with tenurial arrangements in the context of a particular environment/technology/tenure system complex.

Among the desirable attributes of agroforestry technologies are the ability to increase sustainable productivity, augment income stability, increase economic and ecological diversity, and enhance positive interactions among the components of a farming system. To be adoptable by low-resource farmers, agroforestry technologies should also be relatively simple and require a low level of initial investment.

Group members agreed that the distribution of benefits from new agroforestry technologies is as important as the improvements in efficiency which they are capable of achieving. In this connection it is noted that there are at least two types (or levels) of efficiency: simple "technical efficiency" and the wider "social efficiency" of the technology as seen in the context of the workings of the society as a whole.

Another point of general agreement is the observation that technologies which achieve economies of scale in larger production units often lead to a displacement of low-resource farmers. Not only is it more difficult for such farmers to avail themselves of the opportunities presented by the new technologies, but the adoption of such technologies by the wealthier farmers may lead to the reduction or abrogation of the traditional land use rights of tenants and users of common property resources. Driven by narrow criteria of technical efficiency within the most successful production units, such changes may reduce the overall *social efficiency* of resource allocation mechanisms associated with the traditional land use system.

Of particular importance to low-resource farmers, the group noted, is the uncertainty associated with new technology. Technologies which radically alter the existing land use system and increase the ambiguity of traditional tenure rights are likely to be met with justifiable suspicion by those who are unsure of how their traditional rights will be affected. Under conditions of increasing pressure on resources, the survival strategy of low-resource farmers depends upon being able to take advantage of new technologies without letting go of the security of older, time-proven means of livelihood. In general, it was felt that technologies which are capable of gradual adoption on a small-scale, experimental basis are more consistent with the risk-aversive strategies of low-resource farmers than those requiring major initial commitments of land, labor or capital.

It was further agreed that the agroforestry technologies with the best chance of adoption by low-resource farmers are those which offer a rapid return on investment (e.g. short tree rotations or early returns from the annual crop component of mixed tree/crop associations) and/or those involving tree/crop associations that are dynamic and flexible over time (i.e. capable of incremental adoption and modification in accordance with short term needs for cash or subsistence products). The inherent flexibility of the agroforestry approach was noted, but it was also emphasized that any particular technology should be seen as a means to the achievement of socioeconomic objectives and not automatically taken as an end in itself. In this connection, a particular danger to be avoided is the tendency of promising technologies (e.g. alley cropping) to become so fashionable in technological circles as to override due consideration of their actual appropriateness in particular socioeconomic settings.

Much of the discussion of the problems of resource-poor farmers applies also to tenure considerations in agroforestry technology design. Although specific rights (i.e. to plant trees, protect trees, harvest the products of trees, etc.) must always be subjected to detailed analysis, in summarizing its observations on the main technology-tenure interactions, the group found it useful to distinguish between land users with generally "low" vs. generally "high" tenure rights. This led to the following general conclusions:

1) Where the tenure rights of a particular group are low relative to other groups, no agroforestry technology will help a lot.

2) New agroforestry technologies can exacerbate the plight of poor people with low tenure rights.

3) Where tenure rights are undeveloped and no one enjoys greater rights than others (e.g. in undeveloped wastelands), appropriately conceived agroforestry technologies can provide a basis for both improvements in the productivity of land in general and an increase in the productive resources of poor people in particular.

As a general caveat, however, it should be noted that, while certain technologies are more conducive to adoption by poor people, few if any technologies carry an intrinsic guarantee that the new resource opportunities which they create will be enjoyed exclusively by the poor. To secure the benefits of agroforestry innovations for resource-poor farmers will require not only well-designed technologies, but also well designed projects.

Report of the Working Group on Women, Tenure and Agroforestry

Chair: Marilyn Hoskins
Rapporteur: Louise Fortmann

Is gender a relevant variable?

In earlier sessions of the Workshop, the view that women did not need to be addressed as a separate issue was both implicitly and explicitly stated. Some participants felt that the interests of women were already adequately protected by existing social institutions. Others felt that women were not and should not be involved in the labor involved in agroforestry. Still others felt that perceptions of gender differentiation were an artifact of feminist ideology imposed on essentially egalitarian societies. The danger of the differentiation of issues by gender perpetuating a separate and unequal status for women was also expressed. Yet, in most societies there is a differentiation of labor and decision-making by gender within household economies. In many societies there is gender differentiation in access to and control of resources. Such arrangements can lead to very different interests on the part of men and women. In such situations it is not always possible to rely on the "good manners" of the household to protect the interests and status of women.

It is also the case that even the most egalitarian societies may be vulnerable to the effects of interactions with non-egalitarian societies and to the introduction of new technologies and new forms of social organization. The work of Carol Colfer among the Dyaks of East Kalimantan shows very clearly that the introduction of the chainsaw, which women could not use, and the insistence of outsiders on interacting with the society only through the men, put a very severe strain on previously egalitarian social arrangements.

From the viewpoint of the project planner, it is necessary to look for gender differentiation:

1) To avoid project failure due to the failure to consult with the appropriate person: because of gender differentiation, often only the women will know what trees are used for food, what species have desirable fuel characteristics, etc.

2) To avoid destroying a working system: without adequate knowledge a project may clear the land where women gather fuel, fod-

der, food and other minor forest products which are essential to the household economy.

3) To avoid and/or alleviate social stress: awareness of gender distinctions can alert project planners and implementors to the possible adverse effects on women of well-intentioned interventions in both egalitarian and non-egalitarian societies. For example, some projects deliberately hire women because they may be paid at a lower rate. Such "programmed impoverishment" should not be considered acceptable.

Issues concerning women and tenure

Access to land

Women's access to land is often through tenuous, undefined and changing relationships with men, which the men can often terminate at will.

In Latin America women in common-law unions frequently are without land rights, as in some places where the law specifically prohibits inheritance by members or offspring of such unions. In parts of Kenya couples may elope rather than be married traditionally to avoid payment of a substantial brideprice. The wife in such marriages often has no locally recognized land rights even if she has been legally married under statutory law. When polygamy was outlawed in Vietnam, men kept only one wife (often the most beautiful or the youngest), leaving the other wives and their children disenfranchised.

The shift from customary to statutory law may disenfranchise women who were protected under customary law. Land reform has been known to transform matrilineal societies to patrilineal societies. Resettlement programs frequently focus on the family unit, providing women with their own land only if there is excess land available.

Women may "benefit" in a sense from land fragmentation, if it leads to such small parcels that the men migrate, leaving the land to the women. Women's rights to land must be analyzed according to the quality, quantity and location of land available to them.

Youth

Projects or other activities may absorb available land and time, leaving young men unable to raise enough money to pay the brideprice for a wife. Some youth groups may cut fuelwood to raise money. In other cases, young men may migrate to the city.

"Good manners" or the social contract

All societies have implicit and explicit norms and folkways which might be characterized as "good manners." However, the norms of good manners are defined differently by different groups and under stress (rapid population growth, alienation of land, agricultural transformation) such manners may break down. In such cases, those whose rights are made explicit in statutory law will be protected, while those (often women and youth) whose rights are not may be disenfranchised. For example, widows in Burkina Faso who are not young and productive are no longer taken care of. In Latin America, widows in swiddening societies were taken care of and retained their rights to the family land, but such rights are not recognized by outside settlers who take possession of "unoccupied" land.

Labor

Labor is necessary to utilize land. A woman who is tied to the house by domestic maintenance tasks will be less able to utilize a distant piece of land than will a man. Female household heads frequently control less labor than male heads and therefore are less able to utilize land to which they have rights. Thus land tenure rights may have little meaning without the labor with which to implement those rights.

Class distinctions

Women in different socioeconomic classes generally have different access to, or control over, resources and participate in the economy in different ways. Female households and women in squatter households are often particularly disadvantaged. It is necessary to differentiate women by class for policy and project purposes.

The issue is not the physical act of tree planting but decision-making

The physical act of planting a tree (which can be done by anyone with time) is without personal benefit unless it carries with it the right to use and/or control the tree. Too often women are involved in projects as laborers, planting trees they have no right to use. Nor does this kind of tree planting experience necessarily carry over into the household economy. As a rule, the women who have planted the Kenya Green Belt trees, do not plant trees at home.

Local participation

Participation of poor people, both men and women, is necessary for equity and effectiveness in community tree planting projects. Without such participation, the information necessary to deal with the issues identified above will not be forthcoming. We need to develop better techniques for participatory planning. For example, in some societies it is bad manners for women to complain. In such cases, the traditional meeting format will not be effective. Developing alternative techniques will not be easy, but it is essential.

The "iron law of subsistence"

It has frequently been the case that as soon as women's activity becomes profitable, it is taken over by men. Such "project hijacking" at the household level often serves to perpetuate women's status as marginal subsistence producers. It may help to ensure that adequate resources are available to allow both men and women to participate in profitable activities. It may also be necessary to work through men to facilitate women's involvement and to assist in solving their problems.

The meaning of trees

Some woody plants are not defined as "trees." For example, in some areas of Kenya timber species are defined as trees and may be planted only by men. Fuelwood species are not defined as trees and therefore may be planted by women. Other species may be defined as frivolous, e.g. as "children's fruit." Attention to local definitions of trees may facilitate women's participation in agroforestry where it might otherwise be forbidden.

Time

If women are to exercise their rights to trees and land or to be employed in wage labor, they must have the time to do so. Rural women generally carry a heavier workload than men and must spend more time close to the home. Solutions to this problem include provision of day care, "flexitime," and piecework.

Institutions affecting women's involvement in agroforestry

Banking

Credit is often needed for commercial agroforestry. When credit is needed, women often may not have the requisite attributes such as land title. Project planners need to be aware of the need for credit and to make special arrangements for women's credit.

Schools

School-age boys and girls work as equals in the planting and maintenance of trees. This experience provides young women with the skills to plant and maintain their own trees. It may also convince them of the benefits of tree planting and the inappropriateness of customary restrictions on their involvement with trees, motivating them to find ways of working around those restrictions.

Research institutions

National and international research institutions have historically failed to address the problems and needs of women and have failed to consider women in their roles as the producers, managers and users of agricultural and forestry resources and products.

Extension

Extension services are generally male-dominated at both the field and supervisory levels. There is a need to involve women in extension at all levels. Both male and female extension workers need to be aware of women's problems and circumstances. The practice of providing inputs or information to women via their husbands in the hope that it will "trickle over" has proved to be ineffective. Many husbands never pass inputs along or simply get the information wrong. Extension services must deal directly with women. Frequently this will require the hiring and training of female extension workers.

Extension services need to develop a unified approach. Fragmented extension services may lead to household conflict when the husband and wife (wives) are given conflicting advice. Extension services need also to consider the household as a unified production unit rather than concentrating on disciplinary specialities such as agriculture, range management and forestry. Narrow disciplinary approaches may lead an agricultural extension worker to demand the uprooting of trees intercropped with arable crops for fuel, fodder or fertilization purposes.

Training

More creative ideas (including equivalency criteria and ascertaining that entrance requirements are legitimate) are needed for helping women meet entry requirements for training programs. Housing for women must be provided at training institutions. Adequate and relevant employment must be provided for women after training.

More training should be provided for female village leaders and paraprofessionals. The issue of who will meet paraprofessional salaries at the end of the project needs to be resolved.

Training materials should be distributed through multiple channels (churches, women's groups, literacy classes), not just through extension services. The power of the written word should not be underestimated. Many literates have nothing to read except the occasional label or literacy book. Such people could read written material to other villagers.

Agrarian reform

Agrarian reforms often benefit only those classes of people (such as men) who are already eligible to own land, and women frequently are not the beneficiaries of reform programs. Ways must be found to incorporate women, including female household heads, into settlement schemes and other agrarian reform programs. Written contracts should be used as tools for clarifying specific rights and obligations. Statutory law, written contracts and agrarian reform policy may be appropriate arenas in which middle class women's groups can take action to protect the rights of all women.

Research agenda

1) Look into the relationship between the growing of trees and the growing of food crops:

a) Nutritional effects

Hypothesis 1: Nutrition may improve as fuel becomes more readily available to cook food.

Hypothesis 2: Nutrition may worsen as commercially valuable trees replace food crops.

b) Resource management decision-making

How do households decide how to use resources?

How do preferences for resource use differ within the household?

c) When land is switched from food crops to non-food tree crops, who provides the household food and how?

2) Identify agroforestry techniques and tenure arrangements that will promote the enrichment of women's production and gathering niches: home gardens, fallow land, common lands, and open access lands.

3) Identify women's needs and women's products obtained from forests and trees.

4) Identify which kinds of forest and tree products are necessary for women's small household industries, what institutional arrangements, including tenure, are appropriate for meeting those needs and what kinds of forest policies will meet the need of such enterprises for fuelwood and raw materials?

5) Identify those rights and mechanisms that are conducive to renewable as opposed to consumptive use of trees by women.

6) Identify the institutional mechanisms required to provide women (and others) with access to forests and forest products.

7) Identify the institutions and mechanisms that preserve and perpetuate subordinate gender roles for women, particularly as they relate to agroforestry.

8) Identify the strategies and mechanisms through which women (either on their own initiative or through the facilitation of others) have become active participants and decision makers in agroforestry activities.

Agenda for immediate action

1) A systematic and thorough literature review of women's involvement in agroforestry should be undertaken.

2) In all ongoing and future agroforestry research, all data should be disaggregated by gender, age, class and type of production system.

3) In all ongoing and future agroforestry projects, women's tree tenure and land tenure rights should be systematically explored, documented and integrated into project design.

Report of the Working Group on Agroforestry Project Design

Chair: John Bruce
Rapporteur: James Riddell

This report is an attempt to capture in words a discussion and debate that lasted over three hours. At best it can only touch upon the great variety of issues discussed and try to capture the main points of the many subtle arguments made. If a general point can be made it is, perhaps, that the working group saw agroforestry as in many senses a solution in search of a problem; that is, a project. This is related to two aspects: one being that woody plants are multipurpose in peoples' social and economic strategies. The second is that projects must adapt to a bewildering diversity of social and cultural settings, even though agroforestry is a form of rural resource use that is part of most traditional tropical agricultural systems. The challenge is to see to it that the benefits of project interventions outweigh the costs.

In approaching this multifaceted problem, the working group divided the task into a set of major themes. These should be viewed as suggestions for the organization of research topics to be investigated before initiating agroforestry projects, rather than a firm set of prescriptions.

The social and institutional environment of a project

The group first addressed the question of what it is that first gives rise to the idea of an agroforestry project and its objectives. In general, it was agreed that these may be roughly divided into three areas of concern. First is the objective of rural income enhancement. A second focus is fuelwood production and a third is a conservation emphasis. The emphasis which is chosen is related to who it is that proposes the project.

Who, then, perceives the problem? A development idea is often a function of fashion--a diffusion of enthusiasm within the development community. Research discussion, public awareness of a problem and a certain amount of political activity always precede a project idea.

This raises the question of when and where do we seek the input of the rural producers who will plant these woody plants. A bottom-up approach was undoubtedly subscribed to by all the group's members, but it was recognized that the idea must also appeal to the priorities of both government and donors if it is to compete successfully for scarce funds. There is a need, therefore, to integrate national and local priorities into a single project focus.

Most projects are the result of an extant approach to project design that has a life of its own, closely related to the career structures of those in the donor and government agencies. Each agency has its own development strategy that is a product of its own evolutionary history, and projects themselves have a life cycle that has little to do with the maturation rate of woody plants. At the international level, aid agencies may be mandated to work on a government-to-government basis and the very ministries with which they can deal may also be part of an inflexible mandate.

All of this indicates that the institutional and policy frameworks in which projects are formed are largely predetermined by past experience and are bounded by factors that are as much a part of the environment of a project as any physical or biological features of the area.

In regard to the tenure dimension of projects, the group agreed that the institutional and policy environment of a project will determine which categories of land are addressed by the project, i.e. household, community or state lands. In general, it was agreed, that the further one moves from the household level in project planning, the greater the danger of negative impacts on recipients rights in land and landed resources. Examples were cited of elites being able to siphon off a majority of project benefits for themselves. The best way to avoid this is to include all levels--government, donor agency and local people--in project conceptualization. Nevertheless, it was recognized by the group that projects are most often conceived in an environment that is already bounded by prior policy choices.

Project beneficiaries

It was a general consensus of the group that the benefits of a project must be directly perceived by the rural producers who actually plant the trees. Benefits occur in many forms, e.g. fuel and wood products for urban dwellers, long term enhancement of the environment of benefit to society as a whole, etc., but from the standpoint of project success the needs of the rural producers will be paramount. In addition, it was pointed out, the government and donor groups have already been convinced of the potential benefits, or one would not be designing a project in the first place.

The group emphasized the fact that there are differences in the preference for different benefits on the part of local people and elites, government, etc. This should be made explicit and not glossed over.

Land tenure costs will be borne by the local producers. They will not see the rewards of their participation for a number of years, depending on the woody plant selected, and then there may be unanticipated tenurial impacts. An example discussed by the group was the growing of export tree crops. These are introduced for export earnings and not primarily for local benefits. Once they have come into production, they can create tenure inflexibility leaving the local economy unresponsive to changing demands for food products, etc. It was felt that agroforestry could embody a more flexible approach.

Another way in which rural producers are often vulnerable is the small size of their plots. It was suggested by some that where tree planting increased insecurity of holdings, this may be where the project should consider using state or community land.

Participants and participation inducement strategies

The group felt strongly that project planners need to understand the diversity of interests in land and landed resources within groups and between social levels. This information is used for pre-project planning and for the implementation phases. That is, it was felt that special emphasis be placed on gathering information on the kinds of rights, tenurial and access, *de jure* and *de facto* that are held by women, the landless and other social categories that may be overlooked in planning.

In regard to landless populations the discussion made note of at least five categories. The young farmers coming of age and finding no land in their natal community were one. Another was the truly dispossessed that make up a floating proletariat rural labor force. These are often the actual planters of the trees in many projects. These rural inhabitants are different from the squatters who are also often referred to as landless, and their tenure problems are very distinct. There are also squatters who try to establish a claim to land and there are slash and burn farmers who invade forest reserves. Finally, there are the landless sharecroppers who have almost no possibility to ever possess land of their own. Each of these categories of landless have unique land tenure problems; for example, sharecroppers can easily find themselves excluded from the land once the landlord is convinced that tree crops are more profitable than conventional farming.

The working group also discussed the need to understand the tenure rights associated with different social categories in a society: class, caste, ethnic group, etc. A final point raised in regard to this issue was the need to find out who really has access to the commons. Cases were cited where the common lands were in actuality a preserve of the elite.

The general conclusion on this topic might be expressed as a rule that participants should have no fewer rights in land after the project than before. The project may set changes in motion that will change the actual amounts of land held by particular individuals. Not to be overlooked are those populations that may hold access and other tenure rights to land on a seasonal basis, such as pastoralists and fishermen.

Physical location of projects

The group noted that there are two dimensions in regard to where a project is located: technical and political. In general, it was felt, an agroforestry project may have a series of technologies in mind that have important ecological specificity, but that such a technology cannot simply be imposed on the landscape. The range of possible sites will be determined or limited by government, on the one hand. On the other hand, the project will need to understand the social environment into which it is making interventions. Finally, donor groups often have definite ideas about where they want to place a project. The final location should be a balance between the needs of intended beneficiaries, local social realities and government and donor desires.

In projects that are placed on state land and communal holdings it was felt that we need to determine all the uses made of this space so that existing access rights can be studied. It is often the most marginal members of society that have had to depend (often surreptitiously) on these areas and will therefore be most seriously affected by a well intentioned but uninformed project. In other words, one should see if there are important access rights that are not recognized by the government.

It was also suggested that the choice of a location may call into question local institutional arrangements that have been developed to deal with land tenure insecurity. An example was given from the Peruvian Amazon where two groups of Indians joined into a union to jointly preserve their traditional land from encroachment by settlers. There had already been many attempts to break the political will of this spontaneous organization. The territory covered two river basins. When the project was designed it was to be located in only one of these. The project designers were ignorant of the local political developments and were surprised when the populations vigorously rejected the project as an attempt to break their solidarity. In cases like this, any idea of a project on less than the total area may be viewed as a political threat to land.

Extension strategy

It was very strongly felt by the group that there has to be extension for all. In some regions of the world the elite have unequal access to extension services and therefore they are the ones that are able to reap the benefits of new technology. Another point of general agreement was that extension should not be a one-way process. There has to be some way found to build on indigenous knowledge of how to use landed resources and land. An example of a successful model was presented where a recognized village expert (known to all as a good farmer) was actually retained by the extension service as their local representative. This person provided the follow-up on farmers' responses to the new ideas and also served as a person who could explain why some local practices were superior to those being introduced.

A point on which there was a great difference of opinion concerns the social goals of a project. One side maintained that introducing woody plants into a farming system was difficult enough and that it should be kept separate from social goals. These should be dealt with under another project. This was vigorously contested by others who held that every project has serious repercussions in the rural community and that these impacts should be considered from the very beginning.

The group spent a considerable period discussing the issue of nurseries. Arguments were presented pro and con on private *vs.* cooperative *vs.* government nurseries, each of which takes land from a particular sector. Private and cooperative nurseries were said often to be seen as a threat to government programs. One participant stated that it is the role of extension to see that the best species are planted. Examples were given of private firms "pushing-off" onto the farmer the nursery stock they happened to have on hand, even when this was clearly inferior, or indeed, the wrong tree for the purposes the farmer had in mind.

The general conclusion was that this issue will have to be investigated on a site specific basis. The area where there was no disagreement concerned the location of the nurseries relative to the farmer. The consensus was the closer the better.

Land tenure issues in the discussion of extension were a little more subtle and implicit rather than explicit. In general, the choice of a technology for extension will affect all other aspects of the project and land tenure. It was stated, for example, that if teak is chosen, the long maturation time will probably limit it in most cases to state land and some form of tenancy or exclusion. Rubber, on the other hand, will require organizational collectives like estates or settlement schemes. However, the pines, leucaenas and other fast maturing trees were thought to be compatible with many different tenure arrangements.

Choice of institutional linkages and interventions

Various group members felt that there was a need to start with an inventory of all the institutions potentially involved in a project, from the local to the national level. This should be followed up with a study of their existing roles in the project area. An example cited by several participants concerned the varied roles played by religious leaders like the Imam in Islamic communities and by school teachers more generally. In some cases these people were the animators of the acceptance of new ideas, and in other cases they were part of an exploitive local elite.

Others pointed out the need to identify which institutions determine the "proper," culturally approved, uses of land. An example given was of chiefs in Lesotho who will have trees pulled up if they are planted without permission. The chiefs recognize the value of trees and woody plants in their erosion prone area, but they also have the duty of protecting common property holdings.

The need to examine institutional roles extends also to non-government organizations (NGOs). The workshop in general had been high in its praise of these organizations but in our discussions it was pointed out that they are in reality a mixed bag. Some are good, others problematic and others may be well-intentioned but lacking in capacity to achieve project goals. In this same vein, it was generally felt that, ideally, peasant organizations would be preferable. However, the group recognized that this is in reality very difficult from an organizational point of view and may be politically impossible in many countries.

In regard to land tenure, specifically, it was noted that in some parts of the world the legal system makes agroforestry difficult if not impossible. It may be necessary in such cases to request the host country to review its land tenure and tree laws to remove obstructions. Examples from the Sahel and Latin America were discussed. It was suggested that in Latin America there is a need to allow farmers to cut trees they have planted on their own holdings without being required to obtain permission from the national forestry service. Perhaps this could be done by returning trees to the private property part of legal code while leaving national institutions in charge of protection and conservation of national forests. This was an issue discussed many times during the week and it points out the political economy of institutional linkages that are part of any agroforestry project.

Information

The last topic considered was the type of information needed for project design. Unfortunately the group ran out of time before this discussion could be concluded.

However, a general conclusion that emerged throughout our deliberations was the need for intensive and prolonged studies at the field level during the project design phase. We need to be sure we know who make up the social groups that have interests in land.

Another area often overlooked is the experience of prior projects in the area. There is seldom a region that has not had development efforts introduced and today we find that agroforestry projects are often integrated into a larger on-going effort.

It was repeatedly stressed in the group's discussions that there is a need to monitor the impacts of projects at all levels. An example from Nigeria was given where on-farm trials had to be halted because of land tenure problems. The other members of the larger kinship group holding rights in the land on which the trials were located insisted that tree planting be stopped. Other examples involving sharecroppers were given. What is needed is the capacity to redesign a project in midstream, whenever monitoring reveals serious problems.

Finally, it was recommended to the group that we have information on the planning horizon of the individuals involved. This can be relative to age, social position, group membership, etc. Older farmers, for example, may not want to make investments that they will never see the benefits from. As an old farmer told one of the participants "my grandfather never planted a tree for me." On the other hand a chief may see this as a good thing for a man in his position to do for the community.

Plenary Discussion on Crosscutting Issues

Chair: Bjorn Lundgren
Rapporteur: Barbara Grandin

Technology-tenure interactions

The group's first conclusion was: Where tenure rights are low, agroforestry probably won't help a lot. However, in the discussion it was noted that, even if tenure is not secure, if profitability is high people might risk investments in agroforestry despite poor tenure rights. An example was given from Bangladesh, where poverty-stricken producers grabbed at the suggestion of agroforestry despite poor tenure security. The validity of this conclusion also depends on the definition of agroforestry used. If you take it to mean long-term perennial crops, then it probably holds true. But the true definition of agroforestry should be much broader than that. Many agroforestry systems involve existing trees, or short rotation woody plants where benefits can be seen in as little as 2-3 years. This broader definition of agroforestry is preferable, and widens the scope for involvement of other institutions in agroforestry research and development.

The group's second conclusion was: new agroforestry technology can exacerbate the plight of poor producers who have low rights when the project begins. Discussion first centered around the definition of low rights, and the components that make up "low" rights. It was noted that if someone has low rights it implies that, someone else (person, institution or state) has high rights. The Land Tenure Center uses the concept of a "bundle" of rights, which seems to be a useful way of approching the issue. Westerners tend to think of rights to land as all or nothing; we need to be more aware of the different dimensions of rights. Rights may be plotted on a continuum from individual to communal; rights to use might be different from rights to decide or change or distribute. The issue of the possibility of strengthening rights (especially when they appear "fuzzy") was also discussed. In most situations, it is theorized that clarifying these rights would appear to some people as if you were taking away their rights, and political problems would likely ensue. Exceptions would be possible, for example in wastelands, which are unused, and no one cares about the rights to them.

An example was given of how a change in technology for harvesting palm oil trees (which was conceived of as increasing productivity) led to the disenfranchisement of the poor who had been laborers/sharecroppers.

This led to a discussion of types of insecurity of tenure. The first type of insecurity stems from a tenancy arrangement which is clearly temporary, as in the case cited above. The second type arises when the "bundle" of rights is split; this splitting may be permanent, but can lead to confusion when a new agroforestry technology is introduced which could crosscut rights previously differentiated. In some areas tenants can't plant trees, as landlords fear they would try to make a claim on the land. In other areas, tenants would have no interest in planting trees as they would not be allowed to derive the benefit from them.

It was noted that the discussion tended to focus on users, on those with immediate rights. But rights are vested at different levels, up through the nation state. It is useful to look at the history of an area to assess the insecurity that might arise from past government interventions. The issue of public interest *vs.* private property cannot be ignored when considering security of tenure.

The rights of the poor were felt to be particularly important. It is important to look at general resource endowment as well as tenure. The scale of a proposed agroforestry system is important, because if it is not scale-neutral it will eventually impact adversely on the poor. An example was given of large landowners who decrease area cultivated when labor supply is short, rather than moving to tenancy for fear of losing land rights. For poor producers early returns are also important.

This led naturally to a discussion (albeit brief) of the third conclusion: where dominant rights are not established, agroforestry might improve productivity and give leverage to shift more rights to the poor. Using open access areas is one way that the poor could be given more resources. Although research priorities, on one level, would be highly specific to ecological zone and to nation states, it is possible to identify some general social considerations in the formulation of research recommendations. These would include concentration on producers who tend to have insecure rights, poor resource endowments, etc. Although returns to that research and development project might not be as high as to others, many would agree that technical efficiency and productivity are not the only criteria by which to judge a project.

Women, tenure and agroforestry

The relationship between women and poverty was felt to be important; many more female headed households than male headed ones are poor. It was noted that there are cultural and national level differences in how females are treated (with regard to legal rights, tenure, inheritance, ability to make contracts, etc.); that these should be considered, and that it is important to realize that national law and reality are not always congruent. Examples of both the intention of national law to preserve womens rights, and the ignoring of women's rights (e.g. in agrarian reform) were both noted. It was also noted that poverty is not peculiar to females, or female headed households, and the general issue of social justice should not be ignored. It was felt, however, that there is a limit on the extent to which agroforestry should be expected to solve social justice problems. Nevertheless, in planning agroforestry projects, it is important that these issues not be ignored, and that agroforestry not have unintended consequences of worsening the position of women.

Women deserve particular attention, in addition to poor people generally, as they tend to be more disadvantaged as a group. As the articles in the bibliography clearly show, women have very different tenure rights. Failure to recognize these has often worsened womens' positions.

Related to worsening positions is the question of whether and under what circumstances agroforestry might possibly have a negative effect on food supply, since as goods become commercialized they tend to move into the male domain. It was not possible to resolve this issue; clearly it depends on the product and the mode of introduction. However, planners should be aware of the issue. More research is needed.

From the nutritional point of view, it was also noted that a possible strategy would be to identify areas where nutrition is a problem, and to specifically see if some agroforestry interventions could be used to increase the food supply.

A similar approach was suggested for fuelwood supply, but some concern was expressed that institutions might get bogged down in microlevel studies. Nevertheless, it is clear that fuelwood shortages are causing tremendous problems (leading to erosion, changing food habits as women can't cook, etc.). With regard to the commercialization of fuelwood, one participant wondered whether there was evidence of changes in the traditional division of labor with regard to fuel gathering. Evidence generally shows that men collect fuelwood in some areas to sell, in other areas women do the collecting, but men still sell. Evidence from Botswana indicated that where fuelwood was near residential sites, women collected it. As it became depleted locally, boys and girls had the responsibility to collect it (as women were tied more closely to the home area). But where distances were very great and transport required, men began collecting.

It was emphasized in the discussion that it is necessary to know not only who does the work but also who makes the production decisions. It was recognized that much less is known about decision-making than about agricultural labor. This relates in part to the status of women at the family level, i.e. to the authority structure within the household which varies tremendously cross-culturally. No consensus was reached on the extent to which agroforestry projects could or should challenge this status quo. Examples were given of the differential statuses of women both within the household and within the wider community. The same issue arose in both cases: if we fit a project to customary law we may reinforce women's poor statuses. On the other hand, it was noted that women's statuses are often better in customary than in statutory law. Moreover, while it is possible to pressure for change in statutory law it is much harder to induce change in customary law. Some participants felt that customary restrictions (e.g. women not inheriting family property) make sense in that they can prevent fragmentation of holdings, and the passing of holdings to other families. It was noted that customary laws are frequently very complex and that the discussion could only deal with them superficially. It was suggested that if access to resources is by membership, we need a typology of members. Until that is available it is inaccurate to say that rights are or are not permanent.

Although it was agreed that customary systems are very complex, it was noted that it is essential that they be looked at in agroforestry programs. Customary systems are dynamic, but do not necessarily keep up with other changes in the social and economic order. The transition from predominance of customary to predominance of statutory law was noted as being a very difficult one, particularly for women, whose status is most likely to change in that transition. The conclusion from the discussion of the legal aspects was that more information is urgently needed.

In terms of agroforestry, it was noted that in some areas women are excluded from participation in technical training for forestry. The suitability of women as foresters was questioned, but the majority felt that as agroforestry is not heavy timbering, and does not involve heavier exertion than most women engage in agricultural pursuits, there was no reason to exclude women from agroforestry, either on the production level or in scientific/management careers.

Project design

Carrying on from major issues raised in the previous discussion, an important point of concern was the extent to which projects should be used as a tool for social reform, whether social reforms should be made in order to allow projects to work, and the lengths to which project designers and implementers should go to make sure that the project would not worsen anyone's situation (both participants and non-participants).

The working group accepted "project design" as an activity that would continue; they did not consider whether projects should occur at all as a development strategy. To a great extent certain parameters (e.g. government's chosen development pathway) were taken as given and unchangeable.

It was noted that there is often (if not always) a divergence of interests between different groups involved in projects (within the community, between community and government, government and donor, etc.). It is not always feasible to state these openly; hence, opportunities for real compromise are lost.

Flexibility in implementation and redesign were considered to be essential features of well managed projects. Many projects have rigid workplans and fixed timings which curtail flexibility.

Indigenous knowledge was felt to be an important, although often neglected information base for project planning. As traditional agroforestry long predates "scientific" agroforestry it was felt particularly important to tap local knowledge and understand local, traditional agroforestry systems.

There was a lively discussion on the institutionalization of projects. Three models were proposed: 1) totally governmental, 2) totally private (through NGOs or PVOs), and 3) a mix of these two. Most participants agreed that on some level government needed to be involved, if only to sign the memorandum of agreement. The degree of national involvement is a parameter that must be assessed on a case by case basis; it can't be determined in blanket terms. NGOs are not a panacea; there are good and bad, effective and ineffective NGOs.

Given its mandate, the working group focused mainly on formal project design. However, it was agreed that much development takes place outside of formal projects and that there is a tremendous potential for assisting the expansion of spontaneous development. This often neglected area requires more study. People themselves (independent of government and NGOs) change and develop. However, this type of development can have its pains and costs to people as well.

It was suggested that, on the whole, agroforestry seems to be acceptable on a policy level; people can implement agroforestry if they are given the right advice and have confidence in their government (to give reasonble advice, protect interests, etc.). An example was given from the Sudan on local level institutions imposing sanctions for nursery protection.

While the discussion focused on project design, it was also felt that more information was needed on what happens in an area after the termination of a project. A study of the residual effects (both positive and negative) of "failed" projects was felt to be particularly important.

Part 4

Postscript

Tenurial Aspects of Agroforestry: Research Priorities

John Bruce and Louise Fortmann

Land Tenure Center, University of Wisconsin
Department of Forestry & Resource Management, University of California at Berkeley

Research needs

One of the central purposes of the Workshop was to define research needs concerning agroforestry/tenure interactions: to discover what we need to know but do not know, or understand very imperfectly. This postscript to the Workshop seeks to draw on the wealth of insights in the papers that were presented to that end. We have tried to marshall them in two steps, first a set of issues and needs of relevance to the research community at large, fundamental issues concerning tree tenure relationships about which we know too little; and then a set of issues posed with agroforestry projects in mind, questions which project designers and managers need answered if projects are to succeed, but which are often never asked.

Research on the tenurial aspects of agroforestry must be of two kinds: documentary and explanatory. Documentary research would involve inventorying existing systems of land and tree tenure. It is necessary because we know so little about existing systems of land tenure and more particularly tree tenure. Explanatory research is necessary because we know even less about the dynamics of the interrelationships among land tenure, tree tenure, agroforestry and other technologies, ecology and other variables. In particular, we need to know how changes in one variable lead (or do not lead) to changes in the others. In short, we need to know what is out there and what effects our interventions (both witting and unwitting) may have on the ecology and the people as mediated through the tenurial system.

Documentary research

The need for definitions

Current forays into tenurial research are hampered by confusion in basic definitions. What is a tree? The answer to this seemingly simple question has many branches. Silviculturists look with distain upon lay people who refer to bamboo and bananas as trees. In the Pacific Northwest of the United States, foresters consider only conifers trees, referring to the *Quercus* species (which in the Northeast of the same country is generally referred to as the *mighty* oak) as weeds.

The lesson is that if we want to affect people's behavior we must put biological definitions in their proper perspective and concentrate on trees as social constructs. The social definitional rules are myriad. Some define trees as any tall, woody perennial. Others may define trees by their product, regardless of their botanical characteristics. Thus, plants which produce timber are trees; those which produce fuel wood or wild fruit are not. Still others may define trees by their users. Plants used by adults are trees; those used by children are not.

Similar ambiguity may attach to the definition of forest. In Nepal, peasants habitually planted fodder trees at wide spacings, believing that to plant them closer together was to court seizure of the land as a forest by the government (see Panday, this volume). The operating definition in this case was one of density. Others may use species composition as the criterion. Thus, an aggregation of one species is a forest, that of another species is worthless bush.

Clearly, the initial task of any researcher is to determine the local definitions of the relevant terms (always bearing in mind that different groups in the same area may have widely varying definitions); the definitions under statutory law; and the definitions used by government bureaucrats on a day to day basis.

Documenting existing systems of land and tree tenure

Elsewhere in this volume analytical frameworks for both land and tree tenure are provided and will not be repeated here. Suffice it to say that studies have often overlooked the distinction between land and tree tenure. Because these systems have historically differed significantly even within reasonably small areas and because well-documented systems are likely to have changed under various pressures in recent years, it can be misleading to depend on existing documentation. This is particularly critical in the case of tree tenure which has largely been ignored. It may be useful to determine if there are different kinds of rights for renewable and consumptive uses. Historical effects on these systems should be examined. The degree to which the existing system affects species selection should be documented.

Inheritance

Available evidence suggests that very complex inheritance systems exist for trees and rights to trees. This evidence is often dated, however, and present day systems need to be documented, particularly where customary law conflicts with statutory law. Who inherits what under what circumstances?

Special groups

In any examination of tenure, it is necessary to pay special attention to groups at risk. They may be disenfranchised or under particular pressures or suffer from low or minority status. As tenure rights will reflect the distribution of power in a society, generally one should look for class, gender, age, religious and ethnic distinctions. More specifically, it is necessary to determine the established patterns of access to land and trees of women, the landless, youth, shifting cultivators, pastoralists, tenants, sharecroppers and others who have only temporary claims to land. Any of these groups may be excluded from agroforestry projects by their tenurial status or may be disenfranchised (that is, lose existing rights to land and/or trees) by the implementation of a project.

Explanatory research

Security of tenure

Agriculturalists and foresters became aware of land tenure when it became clear that their projects were being affected by it. Since then security of tenure has become a phrase to conjure by. Unfortunately it is one which is used very loosely and which is often colored by political ideology and by a familiarity with systems of tenure limited to those based on current English property law. Thus, many believe that security of tenure can only exist where there is individually held private property to which a clear written legal title has been established. There is considerable literature which demonstrates that adequate security of tenure can be provided under a variety of property regimes, and that individual ownership can in certain circumstances be ecologically and socially destructive. Much of this literature concerns livestock production in semi-arid areas. With the partial exception of swidden production, there has been much less research on production systems involving trees.

The research question is perhaps more complicated for forested areas than for savannah because there tends to be more competition from outsiders for land in forested areas and because their government is likely to have a greater stake in enforcing the statutory system of land tenure in these areas in order to benefit from timber revenues. Thus the system to be examined is likely to be an evolving interaction between the customary system and the statutory system of tenure and to reflect the changing nature of center-periphery relations. The research problem is complicated by the fact that one can have secure tree tenure without secure land tenure and in some situations, secure land tenure without secure tree tenure.

About this constantly shifting ground, the following questions need to be asked: What mechanisms provide informal security of land tenure? It is necessary here to determine how the customary system worked to provide security. Was security of tenure tied, for example, to marriage ties or symbolic kin ties such as godparenthood? Were there community sanctions, formal or informal, against trespass on, or misuse of, the property of another? What were the rights of outsiders? How were outsiders controlled?

Once this base has been established, it will be necessary to determine the dynamics of change and how this has affected tree use and tree planting.

What are the factors that increase or diminish security of land tenure? Factors to be considered include local population increase, immigration by outsiders, national legislation, changes in cropping patterns, and changes in the value of local crops and natural resources including land.

What is the effect of the imposition of a national statutory system of formal land tenure rights? Are tree tenure rights assumed to pass with the land? Will certain groups lose land and/or tree tenure rights? Will a class of landless or near-landless emerge? What are the different effects at different scales of production?

A special problem arises when trees (usually a few highly valued species) are treated similarly to minerals. In such cases concessionaires licensed by the government (and only such concessionaires) may be entitled to harvest such trees wherever they may be growing. Such a policy not only may wreak havoc with people's tree rights but may also induce considerable insecurity; since a felling crew in the middle of an arable field can do considerable damage. Similarly, legislating national ownership of certain trees (or all trees) is likely to be viewed with disfavor by those who thereby lose property rights and to create disincentives for tree planting. Legislative prohibitions of all treecutting without a permit from the relevant government agency, while usually motivated by considerations of conservation, can pose a major disincentive to treeplanting, especially when the relevant agency is not operational or bribes are sought for the grant of permits. Agroforestry requires the harvesting of trees, not that they be treated as national landmarks to be preserved. Conservation policies need careful scrutiny, and conservation measures need to be more narrowly and carefully drawn.

The effects of the existence or lack of security of tenure on tree planting need to be more carefully researched. Conventional wisdom has it that people plant trees only when they have security of land tenure. The point is valid in some situations, but it can mislead in others because it overlooks the relationship between trees and land. In some places the person who plants trees secures the rights to the land on which they are planted. In such cases we may expect to find tree planting as a response to insecurity of tenure. Elsewhere tree cutting demonstrates land ownership. And in some places, the rights of the landowner and the temporary users regarding trees growing or planted on the property are very clearly spelled out in favor of the landowner. In these two cases, we would not expect the insecure tenant, squatter or sharecropper to plant trees.

It is particularly important to examine carefully the effects of individualization of land tenure on tree planting. There is evidence to suggest that individualization and privatization of land tenure may lead to increased tree planting; the argument being that given the lengthy *rotation* for most tree crops, it is necessary to be assured of long-term ownership of the land. On the other hand, individualization has been known to lead to the destruction of tree resources such as sacred groves which many believe serve important ecological functions. Owners of small individual parcels may feel that they must devote every inch to conventional food or cash crops and thus refrain from planting trees. Clearly, the circumstances under which individualization leads to tree planting or destruction of trees must be carefully studied. It is hypothesized that these effects will be influenced by such variables as the size of the individual plots, the nature of the local production system, the economic value of various trees and the security of tree tenure *vis-a-vis* the government.

Individualization may also lead to the cessation of certain tree uses, particularly when common land is individualized. Honey production, for example, has fallen off in places in Kenya where honey trees have been included in individual parcels of land. This appears to have occurred for two reasons. First, in some cases other uses of the land may give the owner a greater economic return. Second, access becomes more problematic under individual ownership particularly if the land surrounding the tree has been planted to crops. It might be expected, where a community has made general use of a limited number of trees for a certain purpose, that production will fall off with individualization unless the legitimacy of communal ownership of the trees is strong enough to overcome the claims of the individual land owner.

Insecurity of tree tenure needs also to be examined. This sometimes is associated with actions by national governments. For example, Forestry Departments have acted as "tree cops" and trees have been nationalized even on private landholdings. Where insecure tree tenure exists we would not expect to find much enthusiasm manifested for planting trees. Similarly governments may sweep through forested areas destroying them (and the labor invested in them) for military purposes. In other cases, individuals who wish to plant trees as a private good may be discouraged by community norms which define trees as common property.

The commons

Conventional wisdom influenced by the writings of Garrett Hardin tends to assume that the commons is a place of tenurial chaos. Again research on pastoral systems has shown this to be untrue. However, the use of trees on the commons and use of the commons for tree planting has been less rigorously researched. The general problem is complicated by the fact that trees, even those growing on private property, may in some systems be considered common property.

Researching tenure on the commons will require a documentation phase. Who has what rights when? Who enforces restrictions? What is done about would-be free riders? What is done about outside marauders? Does this differ according to their political or economic connections? Such research requires considerable care because in some systems this is the arena in which class or caste distinctions are played out to their fullest. Supposedly common land may have in fact been hijacked by powerful families. Or the poor may make their living there in ways which are not perceived or authorized by the state or by local authorities. Or the state may not recognize the community's right to the commons at all, and press a claim of its own.

Once the existing system of use, rights, and regulations has been identified, additional issues must be considered. Does the commons serve as a safety valve for the poor, allowing the state to avoid redistribution of highly skewed land resources? Are there conflicting uses? For example, how do activities such as grazing and tree production coexist? One possibility is to include coppicing fodder trees in the species mix. Means of strengthening and legalizing use rights must be found. Most important, means of regulating use by members of the community and by outsiders must be identified. Should use rights be assigned to individuals, groups or communities? Should those rights attach to particular trees, to the entire resource area, or to particular products or activities?

Interaction effects: ecology and technology

Certain land tenure systems seem to be more appropriate for certain ecologies particularly if social justice is a criterion for judging the system. For example, communal land ownership appears to be particularly rational in semi-arid areas with highly unpredictable rainfall patterns. At least two ecological dichotomies need to be considered: highland/lowland differentials and well-watered/arid and semi-arid differentials. The kinds of land and tree tenure which may be conducive to agroforestry in these areas have not been determined. However, the issue is becoming an increasingly urgent one as the landless from one ecological zone move into a different zone bringing their traditional (and possibly inappropriate) tenure systems with them.

Varying technologies may be fitted into different tenure systems or may require changes in the tenure system. For example, a highly valuable agroforestry crop may have the same effect as mechanization, or drive land tenure towards privatization. Tree planting on boundaries may play an important role in stabilizing holdings, and those boundaries are a narrow but distinct socio-ecological niche. There is a clear need for a better understanding of the relationship between different forms of agroforestry and changes in the tenure system.

Multiple systems of land and tree tenure

As noted above, the questions of land and tree rights are complicated by the simultaneous existence of customary and statutory systems of tenure. Such situations can lead to a great deal of acrimony. The Government may accuse the peasantry of illegally occupying state lands whether by accident or intent. In turn the peasants can accuse the Government (or government officials abusing their power) of illegally expropriating their lands. Ensuing violence is not uncommon.

Research on the effects of such conflicts and the means towards their resolution is of the utmost importance. The failure to understand these effects can have negative consequences. If the government enforces statutory law without consideration of customary law, the result is generally the loss of rights on the part of local people who may as a result suffer serious economic, ecological and social setbacks. Tree rights are particularly subject to loss since it is rare that statutory law includes the same degree of detail about tree rights as is found in customary law. The received law of many countries which were once European colonies only contain, as part of the imported law of property, the presumption that permanent crops such as trees are "fixtures" and as such belong to the owner of the land, a presumption which runs contrary to customary practices in many societies. While tree rights are legislated in some exceptional cases (for example, Indian law codified the rights of tenants to trees on the land they cultivated) attention to tree rights in statutory law tends to involve the transfer of rights to the government, as in the case of the nationalization of trees or many conservation laws. In addition, local or national elites who understand statutory law may manipulate it to their own advantage, taking common land for their own use and hijacking projects to their own benefit.

Forestry and agroforestry policy

In few places has there been a wholesale abandonment of colonial forestry policy. Rather, many present day forest policies reflect colonial policies sometimes even to the extent of utilizing identical regulations and legislation. Even those countries which have started afresh are affected by the legacy of the negative attitudes toward forestry officials created by colonial practices. This interconnection between colonial and present day forest policy and its effects needs to be explored.

Policy needs also to be analyzed in terms of its latent and manifest functions. It has repeatedly been noted that laws forbidding the cutting of trees or nationalizing trees frequently function to discourage their planting as well. Yet governments continue to perpetuate this fundamental error. Forestry laws may well affect species selection. Of particular concern in many areas is the militarization of the forest in those areas where forests have become battlefields in ongoing guerilla wars. More research is needed to compare what forest policies are intended to do with what they actually accomplish.

Suitable units for agroforestry

Since the appropriate spatial and organizational unit for agroforestry will vary with the problem at hand, it is necessary to determine what kinds of tenure rights are associated with what units of land and with what units of social organization. Family fuelwood requirements may, for example, be addressed through family woodlots. Social erosion on an entire watershed, in contrast, will require the coordinated efforts of much larger units and may involve attention to communal land. Research is needed to determine what tenurial arrangements and what social units can most successfully be involved in the solution to different problems.

The agroforestry project context

It is essential that the designers of agroforestry projects think through and enunciate a tenure strategy, just as they plan a marketing strategy or an extension strategy. It should include both land tenure and tree tenure. Because tenure is not very malleable in the short term imposed by most project time horizons, such a strategy will usually involve designing the project to mesh with evolving tenure arrangements in the project area, and only be concerned in a modest way (if at all) with modifying tenure rules. A tenure strategy is critical to project success: tenure arrangements will dramatically affect the distribution of project benefits, and thus the incentive structure of farmers as they decide how to respond to the opportunities posed by the project.

Several intriguing possibilities as regards strategy were raised at the Workshop, including a phased introduction of new practices to minimize tenure-related strains, and the provision of short-cycle technologies for farmers with limited security of tenure. Such a strategy must also take into account the impact that the project will have on tenure patterns.

Research elsewhere will teach project designers and those responsible for answering their questions to frame more relevant and perceptive questions about tenure/tree interactions, but those questions will need to be answered by inquiry at the project site, at an early stage in the project design process. In the following pages we have tried to address research needs in the agroforestry project context in three ways: a preliminary comment on benefits and beneficiaries, a set of key inquiries to be made as part of the preproject research process, and some general thoughts on such research as a component of the project design process.

A preliminary comment on benefits and beneficiaries

A clear and realistic calculation of the benefits to be produced by the project, and an identification of precisely who will be the beneficiaries of the project are absolutely essential to the development of a sound tenure strategy. Agroforestry projects present special complexities in this regard because one of agroforestry's strengths is the multipurpose nature of tree crops. But puffery in project cost-benefit analysis which promises a plethora of benefits to all concerned (enhanced income for all farmers, lower prices for urban consumers, import substitution, increased exports, and improved hard currency position, whatever!) obscures what in implementation turn out to be hard choices. If those choices are not perceived and clear priorities set in the design stage (and good tenure information does much to reveal the need for such choices), then it is extremely difficult to frame a sound tenure strategy. Information on tenurial opportunities and constraints in the project area need to be integrated with tentative preferences as regards such priorities in a process of mutual accommodation in order to produce realistic project designs. Tenure considerations, as suggested below, can affect project design decisions as fundamental as the project location and the choice of species.

Some critical inquiries

Any attempt to generate a general list of critical inquiries is more than a little presumptious and the list is liable to prove less than adequate in a given project situation. But there were so many important points made in the Workshop that we felt we should venture it. We suggest the following, without suggesting that they are in any way exhaustive:

1) *For the various project design decisions, in the particular society, is focusing on land tenure alone adequate, or are there rights in trees*

distinct from rights in land, so that both require serious consideration? As noted earlier, the Western presumption that trees naturally belong to landowners is in many cases misleading in traditional societies. Moreover, the state may have legislated against such a presumption, establishing public rights in trees. Even where separate tree rights are not established, conservation legislation often limits the landowner's rights in trees upon his/her property.

2) *What is the relationship between the planting of trees on land under particular tenures and the incidence of project benefits?* Tenure systems are just that: systems, complexes of tenures, distinctive sets of rights over particular pieces of land defined by land use and the holders' social function, among other considerations. This variety is best seen by project designers as a set of potentials among which they may chose, a "menu," as one Workshop participant aptly put it. There are tenures (niches in the tenure system) whose selection may benefit the landless (e.g. the commons); tenures whose selection may favor an elite (e.g. owned holdings in situations of skewed land distribution, or the commons where the elite control access to it); and tenures whose selection may benefit women (e.g. the household "garden").

3) *Do the intended project beneficiaries have sufficiently secure tenure in land and/or trees in some tenure niche to create the incentives for them to seize the opportunities presented by the project?* When considerable investments of cash or labor are required, or significant opportunity costs are involved, security of tenure will be a critical consideration. Third World farmers are--most rationally--risk-averse.

4) *Are there ways to enhance the security of tenure in land and/or trees (and thereby the incentives) for those to whom the project addresses opportunities?* Tenure change in the project context must usually be modest. Within the time frame of most projects, tenure change must originate at the local, not the national level. But local institutions are in some cases able to respond creatively to new opportunities by changing local tenure arrangements, provided incentives are real and relevant to those who can effect such changes.

5) *How will agroforestry affect the evolution of tenure patterns and access to land?* Agroforestry projects may need in the short term to accommodate themselves to existing (and evolving) tenure arrangements, but in the long run they have great potential for altering both the terms of tenure and the distribution of land and trees. Will tree

planting enhance the land rights of the planter? If so, at whose expense? At the expense of stranger-farmers, at that of tenants, at that of future generations, at that of other ethnic groups with whom use of the land has been shared on a seasonal basis? Alternatively, may an attempt to plant trees result in eviction of the planter from his or her holding? Will selective access to seedlings provide an opportunity for land-grabbing and contribute to an increasingly skewed pattern of land distribution? Special attention must be given, as noted earlier, to less powerful segments of society, with more tenuous holds on resources. Where asking these questions produces alarming answers, project redesign will be necessary. The redesign process should be seen as a creative dialogue between project objectives and the local reality, continuing until the right instrumentalities are found. While such redesign efforts are always exercises in the art of the possible, we want to stress that there are project ideas which, when examined carefully in the light of the local reality, deserve to be abandoned.

6) *What institutional linkages will be most effective in producing the desired project effects?* Even a farm forestry approach, focused on the individual or family farm, must usually deal with a certain degree of community control over land and tree use. (In land matters one almost never deals exclusively with the individual; a community dimension is always present.) That community dimension becomes especially critical whenever a project tenure strategy involves an attempt to induce tenure change. There are different levels of social control to be considered, with institutions at various levels in the social hierarchy holding different powers with respect to the land. A land tenure strategy must identify the local institutions upon which it will need to rely. But even when agroforestry initiatives seem well suited to existing farm tenure patterns, the projects may need land for special purposes, such as nurseries, which require mobilization of community initiative. One may find the authority to administer community land vested in an institution which is also capable of mobilizaing and organizing community effort (for instance, to protect seedlings) or greater competence for such purposes may reside in another institution. There are choices of social scale to be made, and choices between a variety of existing and potential community organizations (traditional institutions, often more effective in land matters, but sometimes hierarchical and exclusive; more broadly-based but often politicized peasant organizations; technocratic and usually less politically awkward NGOs; or the weak "local government" most commonly preferred by national governments).

Research as part of the project design process

Workshop participants were divided as to how singularly to pursue production objectives, or how heavily project design should be weighted toward achievement of distributive objectives. They were one, however, in a concern that such decisions be made consciously and explicitly, based on solid information on tenure.

How do we operationalize the satisfaction of this need for information in the project context? Rapid rural appraisal methods have been developed which will provide an initial reading on tree/tenure interactions. The questions posed in the previous section are most likely to be of value in that diagnostic context. Such rapid assessment in the project area can often be supplemented with information from at least three other sources: dispute studies, to identify trouble spots in the tenure system; prior project experience (with increasing frequency, the project at hand will have been preceeded by other, unsuccessful projects aimed at tree-planting); and the outcome of the on-farm research trials dictated by farming systems research. As regards the last source, those responsible for on-farm trials need to be more aware of the potential for previewing tree/tenure interactions through the experience in the trials.

These inquiries are a first diagnostic phase in project-related research, and the appropriate shape for later research depends upon their outcome. If on the basis of these inquiries the tree/tenure interactions seem to be fairly predictable and unlikely to pose problems for the project or create other negative consequences, it is then necessary to articulate the assumptions about those interactions as a land tenure strategy, and to ensure that they are monitored as the project goes forward. Likewise, if there is a potential problem posed by a tree/tenure interaction which is readily resolvable, the necessary steps should be taken and their impact monitored.

But if the diagnostic inquires seem to suggest the possibility of serious problems arising from tree/tenure interactions, the appropriate response is more intensive research to a) confirm the likelihood of such problems, b) assess their impact in quantitative terms, and c) acquire the broad range of information necessary to allow development of a tenure strategy which deals with them effectively. To be effective, this research must include both in-depth work of an anthropological/rural sociological nature, and sample survey work to produce the data necessary to quantify phenomena. The research product must meld the relatively sophisticated conceptualizations of indigenous tenure developed by some disciplines with other disciplines' ability to quantify, determine statistical significance, and relate tenure issues to development strategy. Here the three to six week consultancy is not an appropriate approach--however adequate it may have been in the diagnostic stage. A six month effort by a multidisciplinary team is a realistic level of research effort in such a case. A local research institution can often do a very credible job, though it may sometimes need the advice and support of an outside institution with extensive experience in this type of research. What needs to be emphasized here is that the research needed is far more complex than a simple documentation and elaboration of the problems noted in the diagnostic phase; it must obtain an understanding of the social and economic dynamics of the society concerned sufficient to devise an effective tenure strategy.

In any case, once a strategy is defined and implemented, a monitoring phase is entered. This relatively low-grade activity seems remarkably difficult for projects to manage well, and monitoring commonly becomes a simple matter of counting seedlings distributed or feet of timber trucked out of the project area. Monitoring tends to be short-term and sporadic. It is suggested that it is possible to have much more effective and sustained monitoring relatively cheaply through one or two routes: attraction to the project of a dissertation researcher in the social sciences who is willing, in return for support from the project, to place the project's concerns on his or her research agenda; or employment of a local research institution, as suggested above. Such research can be interventionist, and involve trials on a limited scale of solutions proposed for the project as a whole. Where such monitoring is conducted, evaluation becomes less difficult and more predictable, and produces more reliable results. All projects are in the end experiments, and we must try to learn something from them.

Conclusion

While there was a strong sense that successful agroforestry projects require social science research throughout all project phases, there was real enthusiasm for agroforestry as an agricultural development strategy, generated in part by its potential for building upon existing patterns of production and tenure. Agroforestry does not require further social science research to establish it as a promising strategy for agricultural development; but agroforestry does need it to realize its full potential.

Registered Workshop Participants

Prof. R.O. Adegboye
Dean, Faculty of Agriculture
University of Ibadan
Ibadan
Nigeria

Prof. S.K. Adeyoju
Head, Dept. of Forest Resource Management
University of Ibadan
Ibadan
Nigeria

Rafael Baraona
Apartado Postal 463A
Cuernavaca, Morelos
Mexico

Prof. George Benneh
Department of Geography
University of Ghana
Legon
Ghana

William Bentley
Winrock International Institute for Agricultural Development
Rosslyn Plaza, Suite 600
1611 N. Kent Street
Arlington, VA 22209
U.S.A.

Dr. Thomas Bonickson
Dept. of Forestry
University of Wisconsin
Madison, WI 53706
U.S.A.

Prof. David Brokensha
Chairman, Environmental Studies Department
University of California
Santa Barbara, CA
U.S.A

Prof. Marion R. Brown
Land Tenure Center
1300 University Avenue
University of Wisconsin
Madison, WI 53706
U.S.A.

Prof. John Bruce
Land Tenure Center
1300 University Avenue
University of Wisconsin
Madison, WI 53706
U.S.A.

Isabel A. de Ceara
Instituto Superior de Agricultura
P.O. Box 166
Santiago
Dominican Republic

Dr. Michael Cernea
Sociology Adviser
The World Bank
Washington D.C., 20433
U.S.A.

Noel Chavangi
KWDP/Beijer Institute
P.O. Box 56212
Nairobi
Kenya

Dr. Luzviminda Cornista
Agrarian Reform Institute
University of the Philippines at Los Banos
College, Laguna 3720
Philippines

Dr. Michael Dove
Winrock International Institute for Agricultural Development
USAID
18, 6th Avenue
Rama-5
Islamabad
Pakistan

Prof. Louise Fortmann
Dept. of Forestry and Resource Management
145 Mulford Hall
University of California
Berkeley, CA 94720
U.S.A.

Dr. Paul Francis
ILCA
c/o International Inst. of Tropical Agriculture
P.O. Box 015
Ibadan
Nigeria

Dr. Barbara Grandin
ILRAD
P.O. Box 46847
Nairobi
Kenya

Dr. Marilyn Hoskins
Forestry Department
Food and Agriculture Organization
Via Delle Termi di Caracalla
00100 Rome
Italy

Prof. Hunud Abia Kadouf
Faculty of Law
University of Khartoum
P.O. Box 321
Khartoum
Sudan

Ted King
Dept. of Forestry
University of Wisconsin
Madison, WI 53706
U.S.A.

Prof. Kalambay Lumpungu
B.P. 204
Kinshasa 11
Zaire

Prof. Gerald Murray
Dept. of Anthropology
University of Florida
Gainesville, FL 32611
U.S.A.

Prof. H.W.O. Okoth-Ogendo
Dean, Faculty of Law
University of Nairobi
P.O. Box 30197
Nairobi
Kenya

Dr. Kk. Panday
ICIMOD
GPO Box 3226
Kathmandu
Nepal

Dr. Komon Pragtong
National Forest & Land Management Division
Royal Forest Department
Bangkok 10900
Thailand

Dr. John B. Raintree
ICRAF
P.O. Box 30677
Nairobi
Kenya

Prof. James Riddell
Land Tenure Center
University of Wisconsin
1300 University Avenue
Madison, WI 53706
U.S.A.

Prof. Bernard Riley
Environmental Studies Department
University of California
Santa Barbara, CA
U.S.A.

Dr. Dianne E. Rocheleau
The Ford Foundation
P.O. Box 41081
Nairobi
Kenya

Prof. Percy E. Sajise
UPLB-PESAM
College, Laguna
Philippines

Dr. Abul Gassim Seif el Din
Sammein Enterprises
P.O. Box 153
Khartoum
Sudan

Dr. Sergio Sepulveda
Animal Production Department
CATIE
Turrialba
Costa Rica

Dr. Gill Shepherd
Overseas Development Council
10 Percy Street
London WC1
U.K.

Dr. G.B. Singh
ICAR
Krishi Bhavan
New Delhi
India

Prof. Loekman Soetrisno
UGM Rural and Regional Studies Center
Bulaksumur
Yogyakarta, DIY
Indonesia

Dr. James Thomson
119 N. Wood Avenue
Silver Spring, MD 20901
U.S.A.

Prof. S.M.P. Tjondronegoro
Dept. of Rural Sociology and Economics
J1. Raya Pajajaran
Bogor
Indonesia

Index

Editor's note: Due to the wide range of technical, legal and socioeconomic terms contained in these proceedings, to facilitate easier access to the material the index has been divided into three subsections: "land use systems," "tenure systems," *and* "other socioeconomic."

Land use systems

Tenure systems

Other socioeconomic

1000 # 93264 Trees
"Land Tenure Center"